SpringerBriefs in Physics

T0202963

For further volumes:
http://www.springer.com/series/8902

Xavier Calmet · Bernard Carr
Elizabeth Winstanley

Quantum Black Holes

Springer

Xavier Calmet
Department of Physics and Astronomy
University of Sussex
Brighton
UK

Elizabeth Winstanley
School of Mathematics and Statistics
University of Sheffield
Sheffield
UK

Bernard Carr
Department of Physics and Astronomy
Queen Mary University of London
London
UK

ISSN 2191-5423   ISSN 2191-5431   (electronic)
ISBN 978-3-642-38938-2   ISBN 978-3-642-38939-9   (eBook)
DOI 10.1007/978-3-642-38939-9
Springer Heidelberg New York Dordrecht London

Library of Congress Control Number: 2013943713

Printed on acid-free paper

Springer is part of Springer Science+Business Media (www.springer.com)

Preface

This Springer Brief emerged from meetings of Working Group 1 (WG1) of the European Cooperation in Science and Technology (COST) action MP0905 "Black Holes in a Violent Universe" which started in 2010. WG1 (chaired by X.C.) is devoted to the study of quantum effects in black hole physics. The term "quantum black holes" (in the title of both the Working Group and this book) refers to black holes for which quantum effects are physically significant. This is a very broad field, ranging from primordial black holes to ones which have mass close to the Planck scale (defined below). Quantum black holes are fascinating objects as they offer a connection between general relativity and quantum mechanics. The most celebrated quantum effect associated with black holes is Hawking radiation. Another important challenge is to understand when black holes form and in particular when the collision of two particles at very high energy will form a black hole. These are some of the raison d'êtres of our Working Group.

Quantum effects are significant for primordial black holes, which could play a very important cosmological role and even be a source of dark matter. Furthermore, one of most important developments in particle physics of the last 20 years has been the realisation that the Planck scale at which quantum gravitational effects become important could be much lower than the traditional value of 10^{19} GeV. Indeed, if there are more than four dimensions in nature, it could potentially be within reach of the CERN Large Hadron Collider (LHC). This idea has triggered significant research which has furthered our understanding of quantum black holes. While so far there is no hint of black holes at the LHC, the data have set some of the tightest limits to date on the Planck scale. Furthermore, the possibility of creating Planckian quantum black holes at colliders was a motivation for ground-breaking theoretical studies of black hole production and decay mechanisms.

In WG1 we have been fortunate to bring together experts in the field of quantum black holes. It was thus the perfect opportunity to review the state of the art of the field. This book is the result of interactions among the members of this working group. Our aim is to have a general introduction to the topic, describing in some detail the theoretical advances that have taken place since the 1970s while being less rigorous in the more speculative parts. In particular, for low-scale quantum gravity black holes, we have not described the current experimental limits in detail as these are evolving constantly as the LHC data grow. Note that we define "quantum black holes" as the ones for which quantum effects are important, although the term is sometimes used more restrictively, to refer to holes for which quantum gravitational effects are important.

We hope that this book will be useful to researchers, advanced undergraduate and graduate students. We assume some familiarity with general relativity, quantum theory and particle physics at the undergraduate level. Throughout this book we use the space-time signature $(-, +, ..., +)$. At various points in this book, we will use relativistic units in which both Newton's gravitational constant G and the speed of light c are equal to unity, as well as electromagnetic Gaussian units in which the Coulomb force constant $(4\pi\varepsilon_0)^{-1} = 1$. In our discussion of black hole thermodynamics we will also set the reduced Planck's constant $\hbar$ and the Boltzmann constant k_B equal to one. Using the constants G, $\hbar$ and c, we can construct a fundamental length-scale $L_P = \sqrt{\hbar G/c^3} \approx 1.6 \times 10^{-35}$ m, mass-scale $M_P = \sqrt{\hbar c/G} \approx 2.2 \times 10^{-8}$ kg $\approx 1.2 \times 10^{19}$ GeV/c^2 and time-scale $t_P = \sqrt{\hbar G/c^5} \approx 5.4 \times 10^{-44}$ s. These are known as the Planck scales and quantities are given in these units when we use relativistic units. When we discuss space–times with more than four dimensions, we denote the number of space–time dimensions by $d = n + 4$, with n being the number of "extra" dimensions. We use the notation $M_\odot$ to denote the mass of the Sun ($\approx 2 \times 10^{30}$ kg), which is a useful mass-scale for astrophysical objects.

Although we had to identify a subset of authors from WG1 to collaborate on this book, we wish to thank all the members of this Working Group. This book would not have been possible without the support of the COST action MP0905. In particular we would like to thank Silke Britzen, the chair of our action, the members of the core group (Antxon Alberdi, Andreas Eckart, Robert Ferdman, Karl-Heinz Mack, Iossif Papadakis, Eduardo Ros, Anthony Rushton, Merja Tornikoski and Ulrike Wyputta in addition to X.C.) and all the members of this action for fascinating meetings and conferences.

We also thank Victor Ambrus, Steven Giddings, Matt Hewitt, Stephen Hsu and Carl Kent for helpful comments on early drafts of this work. Chapter 4 incorporates some material from a recent paper by Bernard Carr, Kazunori Kohri, Yuuiti Sendouda and Jun'ichi Yokoyama. [Physical Review D **81**, 104019 (2010)]. B.C. thanks his collaborators for an enjoyable collaboration and the American Physical Society for authorization to reproduce this material.

The work of X.C. was supported by the STFC grant ST/J000477/1. The work of B.C. was supported by a JSPS/Royal Society bilateral grant and he acknowledges hospitality received from the Research Centre for the Early Universe (RESCEU) in Tokyo University. The work of E.W. was supported by the Lancaster-Manchester-Sheffield Consortium for Fundamental Physics under STFC grant ST/J000418/1.

X.C. is very grateful to his wife Veerle, children Pierre and Aude and parents Jacques and Hélène for their love and support. B.C. expresses appreciation to his wife Mari, whom he sometimes neglected as a result of his work on this book, for her love and understanding. E.W. thanks her parents, Richard Jackson and friends for their continuous support and encouragement.

Brighton, London and Sheffield   Xavier Calmet
April 2013   Bernard Carr
                                                                   Elizabeth Winstanley

Contents

Chapter 1
Introduction

Abstract We discuss the fundamental distinction between quantum and classical black holes. We outline the wide range of ways in which black holes may form, indicating their possible location and formation epochs. Finally, we emphasize the crucial role of black holes in linking microphysics and macrophysics.

1.1 Classical versus Quantum Black Holes

Einstein's general theory of relativity predicts that, if matter is sufficiently compressed, its gravity becomes so strong that it forms a region of space called a black hole from which nothing can escape. The boundary of this region is the black hole's event horizon: objects can fall in but none can come out. In the simplest case, where space has no hidden dimensions, the size of a black hole is directly proportional to its mass. Thus the density to which matter must be squeezed scales as the inverse square of the mass. The Sun would need to be compressed to a radius of 3 km to become a black hole, corresponding to a density of about 10^{19} kg m^{-3}. This is above nuclear density and about the highest density that can be created through gravitational collapse in the present Universe. A body lighter than the Sun resists collapse because it gets stabilized by repulsive quantum forces between subatomic particles.

The Sun itself is not expected to evolve to a black hole but there is a mass range above about $10 \, M_\odot$ in which stars are likely to do so. There should be a billion stellar black holes even in the disc of the our own galaxy, although only the ones in binary systems are readily detectable. Larger "Intermediate Mass Black Holes" (IMBHs) may derive from stars bigger than $100 \, M_\odot$. These are radiation-dominated and undergo an instability during oxygen-burning which leads to complete collapse. The first stars to form in the Universe may have been in this mass range and their remnants may be associated with gamma-ray bursts. IMBHs may also form in the nuclei of globular clusters from the coalescence of smaller black holes. Even larger "Supermassive Black Holes" (SMBHs), with masses in the range 10^6 to $10^9 \, M_\odot$, are thought to exist in galactic nuclei, although their origin remains uncertain. Our own galaxy harbours a $4 \times 10^6\,M_\odot$ black hole and quasars, which represent an earlier evolutionary phase of galaxies, are thought to be powered by $10^8\,M_\odot$ black holes.

All these black holes might be described as "macroscopic" since they are larger than a kilometre in radius. Their signatures might be regarded as astrophysical and they are classical in the sense that quantum effects can be neglected. Although they are not the main focus of this book, we discuss their mathematical and physical characteristics in more detail in Chap. 2.

In the early 1970s it was realized that there are also mechanisms for generating black holes in the early universe. These are termed "Primordial Black Holes" (PBHs) and could be much smaller than stellar black holes. This is because the density of the Universe was much higher at early epochs, reaching nuclear density just 1 ms after the Big Bang and rising indefinitely at still earlier times. PBHs forming sufficiently late can still be regarded as "macroscopic". Indeed, those forming just 1 s after the Big Bang would have a mass of $10^5\,M_\odot$, which is in the SMBH range. Those smaller than 10^{22} kg (about the mass of the Moon) might be regarded as "microscopic", in the sense that they are smaller than a micron, but could have interesting astrophysical consequences. For example, they could collide with the Earth or have detectable lensing and dynamical effects or even provide the dark matter.

Black holes lighter than 10^{12} kg (about the mass of a mountain) would be smaller than a proton and their consequences would be dramatically different. This is because in 1974 Hawking discovered that black holes do not just swallow particles but also emit them. He found that a black hole radiates thermally with a temperature inversely proportional to its mass. For a solar-mass black hole, the temperature is around 10^{-6} K, which is negligible. But for a black hole of mass 10^{12} kg, it is 10^{12} K, which is hot enough to emit both massless particles, such as photons, and light ones, such as electrons and positrons. More details on Hawking radiation can be found in Chap. 3.

Because the emission carries off energy, the mass of the black hole decreases, so it is unstable. As it shrinks, it gets steadily hotter, emitting increasingly energetic particles and shrinking ever faster. When it gets down to a mass of about 10^6 kg, the evaporation becomes explosive, the energy of a million megaton nuclear bomb being released in the last second. The time for a black hole to evaporate completely is proportional to the cube of its initial mass. For a solar-mass hole, this is an unobservably long 10^{64} yr. For a 10^{12} kg black hole, it is 10^{10} yr—about the present age of the universe. Thus PBHs with this mass would be completing their evaporation today and smaller ones would have evaporated at an earlier cosmological epoch. We will therefore classify PBHs smaller than 10^{12} kg as "quantum black holes" since these are the ones for which quantum effects are important. Their influence is more particle-physical than astrophysical and their consequences will be discussed in detail in Chap. 4.

Hawking's work was a tremendous conceptual advance because it unified three previously disparate areas of physics: general relativity, quantum theory and thermodynamics. However, it was only a first step toward a full quantum theory of gravity. This is because Hawking's analysis—and indeed all laws of classical physics—must break down when the density reaches the Planck value of about

$10^{97}\,\mathrm{kg\,m^{-3}} \sim 10^{76}\,\mathrm{GeV^{-4}}$ since gravity then becomes so strong that quantum-mechanical fluctuations in space-time metric become important. An evaporating black hole reaches this density when it gets down to the Planck scale (i.e. a size of 10^{-35} m and a mass of 10^{-8} kg). We describe this as a "Planckian quantum black hole". It is smaller in size but much more massive than an elementary particle. One might expect this to be the smallest possible black hole because space cannot be treated as a continuum below the Planck length. Note that PBHs might form prolifically at the Planck epoch from quantum fluctuations in the metric, although they would evaporate on the same timescale.

A theory of quantum gravity would be required to understand the formation and evaporation Planckian quantum black holes. This might even allow stable Planck-mass relics. However, recently it has been realized that another factor may come into play as a black hole shrinks towards the Planck scale: the existence of extra dimensions. The unification of all the forces which operate in the universe could require the existence of extra "internal" dimensions beyond the four dimensions of spacetime. This approach was pioneered in the 1920s by Kaluza and Klein, who showed that a fifth dimension can provide a unified description of gravity and electromagnetism, providing it is wrapped up very small. Subsequently it was discovered that there are other subatomic interactions and these can be explained by invoking yet more wrapped-up dimensions. For example, superstring theory suggest there are six and M-theory (which unifies the different versions of superstring theory) suggests there are seven.

The extra dimensions are usually assumed to be compactified on the Planck scale, in which case their effects are unimportant for black holes heavier than the Planck mass. However, they are much larger than the Planck length in some models and this has that striking consequence that gravity, which should propagate in these extra dimensions, should grow much stronger at short distances than implied by the Newtonian inverse-square law. In other models, there are different configurations of extra dimensions, known as "warped compactifications", but these have the same gravity-magnifying effect.

This enhanced growth of the strength of gravity means that the standard estimate of the Planck energy (and hence the minimum mass of a black hole) could be too high. This has the dramatic implication that black holes could be made in accelerators, such as the Large Hadron Collider (LHC) at CERN and the Tevatron at Fermilab. These machines accelerate subatomic particles, such as protons, to velocities very close to the speed of light, so that they have enormous kinetic energies. At the LHC, a proton will reach an energy of roughly 7 TeV, which is equivalent to a mass of 10^{-25} kg. When two such particles collide at close range, their energy is concentrated into a tiny region of space, so one might wonder whether the colliding particles occasionally get close enough to form a black hole.

Unless there are large extra dimensions, the likelihood of this is very small because a mass of 10^{-25} kg is much less than the Planck mass of 10^{-8} kg. One can understand this in simple quantum mechanical terms. The uncertainty principle implies that the accelerated particles are smeared out over a distance that decreases with increasing energy and is about 10^{-19} m at LHC energies. So this is the smallest region into which the proton's energy can be packed and it corresponds to a density of 10^{30} kg m^{-3}. But this is well below the Planck density required to create a Planckian quantum black hole. Indeed, in the standard picture a proton would need to be accelerated to the Planck energy of 10^{19} GeV to form a black hole, which is a factor of 10^{15} beyond the reach of the LHC. However, if there are large extra dimensions, we have seen that the Planck scale is lowered and the energy required to create black holes could lie within the LHC range after all, with the holes produced being comparable in size to elementary particles. They would also evaporate shortly after formation, lighting up the particle detectors like Christmas trees.

Although there is still no experimental evidence for this, it opens up the exciting prospect of probing the quantum gravity scale using black holes. In so doing, one could obtain clues on how space-time is woven together and whether it has unseen higher dimensions. A proper understanding of this process requires a careful study of the production of black holes in the collisions of particles, which we present in Chap. 5, while the application to accelerator holes is covered in Chap. 6.

A full theory of quantum gravity will also be required to resolve a profound paradox opened up by Hawking's discovery and one that aims at the heart of why general relativity and quantum mechanics are so hard to reconcile. According to relativity theory, information about what falls into a black hole is forever lost. If the black hole evaporates, however, it invites the question of what happens to the information contained within. Hawking suggested that black holes completely evaporate, destroying the information and contradicting the tenets of quantum mechanics. Various resolutions have been proposed, including the possibility that evaporating black holes leave behind stable remnants, which preserve the original information. We do not discuss this problem in depth in the present work.

1.2 How, When and Where Black Holes Form

In order to supplement the above discussion, we now outline some of the mechanisms for black hole formation in more detail. The important points are summarized in Fig. 1.1. We also discuss the epoch at which the various kinds of black hole might form and their likely location.

MO remnants. The most plausible mechanism for black hole formation invokes the collapse of stars which have completed their nuclear burning. However, this only happens for stars massive enough to be classified as "Massive Objects" (MOs). Stars smaller than 4 $M_\odot$ evolve into white dwarfs because the collapse of their remnants can be halted by electron degeneracy pressure, while stars in the mass range 4–8 $M_\odot$ may explode due to degenerate carbon ignition (Arnett 1969). Stars larger than 8 $M_\odot$ but smaller than about 10^2 $M_\odot$ probably burn stably until they form an iron/nickel core, at which point no more energy can be released by nuclear reactions and so the core collapses (Woosley and Weaver 1986). If the collapse can be halted by neutron degeneracy pressure, a neutron star will form and a reflected hydrodynamic shock may then eject the envelope of the star, giving rise to a type II supernova. If the core

Name	Mass/$M_\odot$	BH formation mechanism/stellar end-state	Location
SMO	10^5	Collapse to SMBH before H-burning from GR instability. Accretion onto/merger of IMBHs	Quasars/galactic nuclei Intergalactic/halo DM?
VMO	200	Collapse to IMBH due to electron-positron instability during O-burning	GCs/ULXs/GRBs? Intergalactic/halo DM?
	100	Explode due to electron-positron instability during O-burning (no remnant)	
MO	25	Prompt core collapse after nuclear burning or delayed collapse after failed supernova	Galactic discs
	1	White dwarf or neutron star remnant or total disruption at carbon ignition (no remnant)	
PBH	10^{15} g	Collapse from primordial fluctuations or at cosmological phase transition or in dust era	Intergalactic/halo DM?
	10^{-5} g	Same formation mechanism but evaporation completed and M_* holes exploding today	Local GRBs?
		Planck mass relics of larger evaporated holes or remnants of Planck epoch	Intergalactic/halo DM?

Fig. 1.1 Summary of various black hole formation mechanisms and their possible locations. *DM* dark matter; *GC* globalar cluster; *ULX* ultra-luminous X-ray source; *GRB* gamma-ray burst

is too large, however, it necessarily collapses to a black hole. Above $40 M_\odot$ the core collapses directly but for 25–40 $M_\odot$ collapse is delayed and occurs due to fallback of ejected material (MacFadyen et al. 2001). Stellar holes must certainly pervade the discs of spiral galaxies.

VMO remnants. Stars larger than $10^2 M_\odot$ are radiation-dominated and therefore unstable to nuclear-energised pulsations during their hydrogen and helium burning phases. However, the pulsations are expected to be dissipated as a result of shock formation and this could quench the mass loss enough for these "Very Massive Objects" (VMOs) to survive for their main-sequence time (which is just a few million years). However, VMOs encounter a serious instability when they commence oxygen-core burning because the temperature attained in this phase is high enough to generate electron–positron pairs (Fowler and Hoyle 1964). This instability has the consequence that sufficiently large cores collapse to IMBHs. Both analytical (Bond et al. 1984) and numerical (Woosley et al. 1982) calculations indicate that this happens for VMOs of mass above about $200 M_\odot$. Various arguments suggest that there may have been a first generation of pregalatic stars in the VMO mass range, in which case they could also be spread throughout intergalactic space and might be associated with gamma-ray bursts. At one stage it was even proposed that pregalactic IMBHs could provide the dark matter, but this seems unlikely since the precursors would generate too much background light (Carr et al. 1984). IMBHs might explain ultra-luminous X-ray sources and those without VMO precursors, formed through the coalescence or accretion of smaller black holes, might reside in the nuclei of globular clusters.

SMO remnants. Stars in the mass range above $10^5 M_\odot$ are unstable to general relativistic instabilities. Such "Supermassive Objects" (SMOs) may collapse directly to black holes without any nuclear burning at all (Fowler 1966). The evidence for supermassive black holes (SMBHs) is now compelling and it is clear that most large galaxies host one (Kormendy and Richstone 1995). Although their formation mechanism is not as well understood as it is for stellar black holes, they could plausibly derive from relaxation processes at the centres of dense star clusters or from the coalescence of smaller holes or from accretion onto a single hole of more modest mass. SMBHs in galactic nuclei could only have a tiny cosmological density. It has also been suggested that there could be an intergalactic population with a much larger density which contributes to the dark matter. However, there are strong dynamical, lensing and accretion constraints on such a population (Mack et al. 2007). Note that the formation of SMBHs does not involve the extreme compression which arises in stellar collapse: an object of 10^9 $M_\odot$ would only have the density of water on falling inside its event horizon.

PBHs. The formation of black holes smaller than a solar mass would require extremely high compression but we have seen that such conditions may have arisen naturally in the first few moments of the Big Bang (Hawking 1971). As discussed in Chap. 4, such PBHs could have formed either from primordial inhomogeneities or from some sort of cosmological phase transition. Since they are expected to have a size of order the particle horizon at their formation epoch, they could span an enormous mass range: from 10^{-5} g for those forming at the Planck time to 10^5 $M_\odot$ for those forming at 1 s. They could even be a population in the IMBH mass range, in which case they might contribute to the dark matter and produce an interesting gravitational wave background without generating too much background light (Saito and Yokohama 2009). If evaporating black holes leave stable Planck mass relics, these could also contribute to the dark matter (MacGibbon 1987).

1.3 Black Holes as the Link Between Macrophysics and Microphysics

The crucial role of black holes in linking macrophysics and astrophysics is summarized in Fig. 1.2. This shows the Cosmic Uroborus (the snake eating its own tail), with the various scales of structure in the universe indicated along the side. It can be regarded as a sort of "clock" in which the scale changes by a factor of 10 for each minute—from the Planck scale (10^{-35} m) at the top left to the scale of the observable Universe (10^{25} m) at the top right. In between are the plethora of fundamental particles, nuclei, atoms, molecules and cells (in the micro domain on the left), mountains, planets, stars and galaxies (in the macro domain on the right), and humans (at the bottom). The head meets the tail at the Big Bang because at the largest cosmological distances, one is peering back to an epoch when the Universe was very small, so the very large meets the very small there. There might also be extra spatial dimensions at the top of the Uroborus, reflecting the higher dimensionality of the early universe.

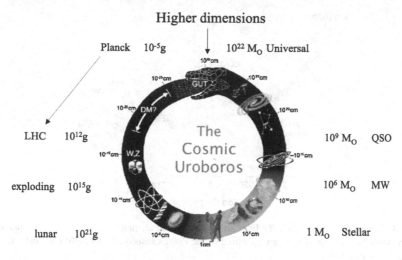

Fig. 1.2 The Cosmic Uroboros: black holes as a link between macro and micro physics. *QSO* Quasi-Stellar Object; *MW* Milky Way; *IMBH* Intermediate Mass Black Hole and *LHC* Large Hadron Collider. Central picture from Abrams and Primack 2011

The various types of black holes discussed above are indicated in Fig. 1.2 and labelled by their mass, this being proportional to their size if there are three spatial dimensions. On the right are the well established astrophysical black holes: the remnants of ordinary or very massive stars, supermassive black holes such as reside at the centre of the Milky Way or power quasars, and in some sense the Universe itself. On the left—and possibly extending somewhat to the right—are the more speculative primordial black holes, which could span the range from Planck relics to black holes evaporating at the present epoch to the sort of lunar-mass black holes which might provide the dark matter. If the extra spatial dimensions at the top of the Uroborus are "large" rather than just having the Planck scale, we have seen that the quantum gravity scale might be reduced to the LHC scale, in which case black holes might be produced in accelerators (10^{12} g). These are not themselves primordial but this would have important implications for PBH formation.

Since our primary focus in this book is "quantum" black holes (i.e. those smaller than the mass 10^{15} g for which quantum effects are important), it is interesting to clarify the link between quantum theory, black holes and quantum gravity more precisely. This involves two key ideas. (1) Quantum mechanics implies that the uncertainty in the position and momentum of a particle must satisfy $\Delta x > \hbar/\Delta p$. Since the momentum of a particle of mass M is bounded by Mc, an immediate implication is that one cannot localize a particle of mass M on a scale less $\hbar/(Mc)$. An important role is therefore played by the Compton wavelength, $R_C = \hbar/(Mc)$. In the (M, R) diagram of Fig. 1.3, the region corresponding to $R < R_C$ might be regarded as the "quantum domain" in the sense that the classical description breaks down there and quantum field theory applies. (2) General relativity implies that a spherically symmetric

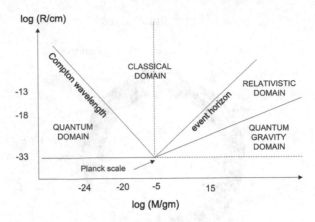

Fig. 1.3 The division of the (M, R) diagram into the classical, quantum, relativistic and quantum gravity domains. The boundaries are specified by the Planck density, the Compton wavelength and the Schwarzschild radius

object of mass M forms a black hole if it is smaller than its Schwarzschild radius, $R_S = 2GM/c^2$. The region $R < R_S$ might be regarded as the "relativistic domain" in the sense that there is no stable classical configuration in this part of Fig. 1.3. The properties of such classical black holes will be discussed in detail in Chap. 2.

The Compton and Schwarzschild boundaries intersect at around the Planck scales,

$$R_P = \sqrt{\hbar G/c^3} \sim 10^{-33}\, \text{cm}, \quad M_P = \sqrt{\hbar c/G} \sim 10^{-5}\, \text{g} \sim 10^{19}\, \text{GeV},$$

and they divide the (M, R) diagram in Fig. 1.3 into three regimes (quantum, relativistic, classical). The vertical line $M = M_P$ marks the division between elementary particles ($M < M_P$) and black holes ($M > M_P$), because one usually requires a black hole to be larger than its own Compton wavelength. The horizontal line $R = R_P$ is significant because quantum fluctuations in the metric should become important below this. Quantum gravity effects should also be important whenever the density exceeds the Planck density, corresponding to the sorts of curvature singularities associated with the big bang or the centres of black holes. This implies $R < (M/M_P)^{1/3} R_P$, which is well above the $R = R_P$ line in Fig. 1.3 for $M \gg M_P$. So one might regard the combination of this line and the $R = R_P$ line as specifying the boundary of the "quantum gravity" domain, as indicated by the shaded region in Fig. 1.3. If there are extra spatial dimensions, the black hole boundary in Fig. 1.3 becomes shallower when M falls below some critical value, so the intersect with the Compton line moves up and to the left. This is why the Planck scales themselves change in this model.

1.4 Conclusions

In this chapter we have seen that black holes form a crucial link between macrophysics and microphysics. Of particular relevance for this book are the black holes for which quantum effects are important and so we will focus on the ones which may form either in the early Universe or in high-energy particle collisions. At the time of writing there is no definite evidence for either of these and it is possible that there never will be. Primordial black holes may not have formed and nature may exclude the production of accelerator black holes. Nonetheless their study is important because they open a unique window onto aspects of fundamental physics.

References

Abrams, N.E., Primack, J.R.: The New Universe and the Human Future, Yale University Press (2011)

Arnett, W.D.: Astrophys. Space Sci. **5**, 180–212 (1969)

Bond, J.R., Arnett, W.D., Carr, B.J.: Astrophys. J. **280**, 825–847 (1984)

Carr, B.J., Bond, J.R., Arnett, W.D.: Astrophys. J. **277**, 445–469 (1984)

Fowler, W., Hoyle, F.: Astrophys. J. Suppl. **9**, 201–320 (1964)

Fowler, W.: Astrophys. J. **144**, 180–200 (1966)

Hawking, S.W.: Mon. Not. Roy. Astron. Soc. **152**, 75–78 (1971)

Hawking, S.W.: Nature **248**, 30–31 (1974)

Kormendy, J., Richstone, D.: Ann. Rev. Astron. Astrophys. **33**, 581–624 (1995)

MacGibbon, J.: Nature **329**, 308–309 (1987)

MacFadyen, A.I., Woosley, S.E., Heger, A.: Astrophys. J. **550, 410–425 (2001)**

Mack, K.J., Ostriker, J.P., Ricotti, M.: Astrophys. J. **665**, 1277–1287 (2007)

Saito, R., Yokohama, J.: Phys. Rev. Lett. **102**, 161101 (2009)

Woosley, S.E., Weaver, T.A.: In: Rees, M.J., Stoneham, R.J. (eds.) Supernovae: A Survey of Current Research, p. 79. Reidel, Dordrecht (1982)

Woosley, S.E., Weaver, T.A.: Ann. Rev. Astron. Astrophys. **24**, 205–253 (1986)

Chapter 2
Black Holes in General Relativity

Abstract As a prelude to our study of quantum black holes, in this chapter we briefly review some of the key features of black holes in general relativity. We focus on specific examples of black hole metrics in four and higher space-time dimensions, which will be needed later in the book. The chapter closes with the laws of black hole mechanics, which raise a curious analogy with the laws of thermodynamics.

2.1 Introduction

Before we can discuss the quantum physics of black holes, we first need to describe their classical behaviour, that is, as objects in Einstein's theory of general relativity. General relativity describes the force of gravity as the curvature of the fabric of space-time. This curvature is caused by matter (or energy, the two being equivalent in relativity due to Einstein's famous equation $E = mc^2$) and is governed by Einstein's equations of general relativity:

$$G_{\mu\nu} = 8\pi T_{\mu\nu}, \tag{2.1}$$

where the Einstein tensor $G_{\mu\nu}$ describes the space-time curvature and the stress-energy tensor $T_{\mu\nu}$ the matter or energy. The curvature of space-time in turn affects the paths of particles, which is particularly relevant for black holes.

In this chapter we provide a brief overview of some of the features of black holes in general relativity, focusing on those aspects needed for later chapters. Further details of black holes in general relativity can be found in the many excellent books available, see for example the classics (Hawking and Ellis 1975; Misner et al. 1973; Wald 1984), modern textbooks (Carroll 2003; Hartle 2002; Hobson et al. 2006) and works specifically on black holes (Frolov and Novikov 1998; Frolov and Zelnikov 2011; Raine and Thomas 2005; Townsend 1997). We begin with descriptions of black holes in four space-time dimensions before discussing some examples of

X. Calmet et al., *Quantum Black Holes*, SpringerBriefs in Physics,
DOI: 10.1007/978-3-642-38939-9_2, © The Author(s) 2014

higher-dimensional black holes. The chapter closes by considering four laws of black hole mechanics, which reveal an intriguing similarity to the laws of thermodynamics.

2.2 Four-dimensional Black Holes

We begin by briefly reviewing black hole solutions of four-dimensional general relativity. Standard theorems (see, for example, Hawking and Ellis (1975)) govern the properties of four-dimensional black hole solutions of general relativity in either a vacuum or coupled to an electromagnetic field. In particular, such black holes are either *static* and *spherically symmetric*, or *rotating* and *axisymmetric*.

2.2.1 Schwarzschild Black Hole

The very first example of a black hole metric (although the nature of the metric as describing a black hole was not understood for several decades) was found by Schwarzschild not long after the publication of Einstein's theory of general relativity (Schwarzschild 1916). The Schwarzschild metric is given by:

$$ds^2 = -\left(1 - \frac{2M}{r}\right) dt^2 + \left(1 - \frac{2M}{r}\right)^{-1} dr^2 + r^2 \, d\Omega_2^2, \qquad (2.2)$$

where

$$d\Omega_2^2 = d\theta^2 + \sin^2\theta \, d\varphi^2 \qquad (2.3)$$

is the metric on the 2-sphere. Here (r, θ, φ) are the usual spherical polar co-ordinates. In (2.2), the constant $M > 0$ is the mass of the black hole. The metric (2.2) is a solution of Einstein's Equations (2.1) in the absence of matter, so that the stress-energy tensor $T_{\mu\nu} = 0$. The metric (2.2) is static and spherically symmetric. As $r \to \infty$, the metric (2.2) approaches the flat-space Minkowski metric, and, accordingly, the Schwarzschild black hole is *asymptotically flat*.

From (2.2), it can be seen that the Schwarzschild metric becomes singular when $r = r_H = 2M$ and when $r = 0$. The apparent singularity at $r = 2M$ is a co-ordinate singularity which can be removed by changing to an alternative co-ordinate system. However there is a *curvature singularity* at $r = 0$ which cannot be removed by a change of co-ordinates.

The curvature of space-time represented by the metric (2.2) modifies the paths of particles. Since, in relativity, particles cannot travel faster than light, we consider the directions of light rays. If a flash of light is emitted from a point in flat space, the light rays will travel outwards from that point in all directions equally. The light rays form what is known as the *light cone* at that point. Since a particle cannot travel

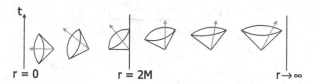

Fig. 2.1 Light cones in the Schwarzschild geometry. The co-ordinate r is the horizontal axis, and t is the vertical axis. The cones at each point are the directions of light rays emanating from that point. A possible particle path at each point is shown in *grey*. Particle paths must lie inside the light cone at each point because particles cannot travel faster than light

faster than light, its path must lie inside the light cone at each point. On the right of Fig. 2.1, far from the black hole, space-time is flat and the light-cones are at 45° degrees to the vertical. As we move from right to left in the diagram, it can be seen that the light cones tilt over due to the curvature of the space-time. At each point in Fig. 2.1, we have shown in grey a possible particle path, which must lie inside the light-cone. Inside $r = 2M$ we see that the light cones have tilted over sufficiently that particle paths must be directed towards the left, that is, towards the black hole, and it is not possible for a particle or light ray to move to the right away from the black hole. The surface $r = 2M$ is the *event horizon*, which forms the boundary of the black hole. The length $r_s = 2M$ is known as the *Schwarzschild radius* of the black hole.

In order to show the complete space-time structure, a Penrose diagram is useful. To draw the Penrose diagram, a conformal transformation is made which brings infinity ($r \to \infty$) into a finite region. The details of the construction can be found in standard books on the subject, such as Carroll (2003), Hawking and Ellis (1975), Raine and Thomas (2005). The Penrose diagram for Schwarzschild space-time is shown in Fig. 2.2. In a Penrose diagram, light rays travel at 45° to the horizontal, and time-like particle paths must have a slope of more than 45° to the horizontal. There are a number of important features of the Penrose diagram in Fig. 2.2. Firstly, the curvature singularity at $r = 0$ has become two space-like surfaces, depicted as the two horizontal lines at the top and bottom of the diagram. Secondly, the surfaces $r = 2M$ are lines at 45° which pass through the centre of the diagram. Thirdly, in the Penrose diagram, the complex nature of infinity can be clearly seen. Infinity ($r \to \infty$) consists of various points and surfaces: the points $i^{\pm}$ are future/past time-like infinity (through which particle paths ultimately pass); the points i^0 are space-like infinity; and the surfaces $I^{\pm}$ are future/past null infinity (through which light rays ultimately pass).

The nature of the surface at $r = 2M$ is evident from the Penrose diagram. Since the surface at $r = 2M$ is at 45°, any light ray or particle which enters into region 2 on the diagram must hit the singularity at $r = 0$ and cannot escape out to region 1. Region 1 is the exterior of the black hole and is the region of interest to external observers. Region 2 is the interior of the black hole. Region 3 is a time-reverse of a black hole, known as a *white hole*, which we do not consider further. Region 4 is a copy of region 1, with another asymptotic region. For black holes formed by matter

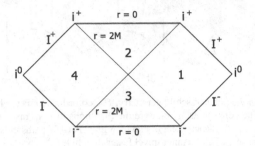

Fig. 2.2 Penrose diagram for Schwarzschild space-time. Time is on the vertical axis, increasing from the bottom to the top of the diagram. Each point on the diagram corresponds to a 2-sphere. The curvature singularity $r = 0$ takes the form of the two horizontal lines at the top and bottom of the diagram. The event horizon $r = 2M$ corresponds to the two lines at 45° passing through the centre of the diagram. The points and lines denoted $i^\pm$, i^0 and $I^\pm$ form infinity

collapsing under gravity, regions 3 and 4 are unphysical (see Fig. 3.1 in Chap. 3). For the remainder of this book, we are interested in region 1 of all black hole space-times, which is the region outside the event horizon.

2.2.2 Reissner-Nordström Black Hole

In the previous subsection we studied the metric of a static, spherically symmetric black hole in a vacuum. It is possible to couple a black hole to an electromagnetic field, which modifies the space-time geometry. The stress-energy tensor corresponding to an electromagnetic field with field strength tensor $F_{\mu\nu}$ is, in Gaussian units,

$$T_{\mu\nu}^{\text{EM}} = \frac{1}{4\pi} \left(F_{\mu\lambda} F_\nu{}^\lambda - \frac{1}{4} g_{\mu\nu} F_{\lambda\sigma} F^{\lambda\sigma} \right). \tag{2.4}$$

Consider an electric field where the only non-zero component of the field strength is

$$F_{rt} = \frac{Q}{r^2}, \tag{2.5}$$

where Q is the electric charge. In this case the solution of Einstein's equations representing a static, spherically symmetric black hole takes the Reissner-Nordström form (Nordström 1918; Reissner 1916)

$$ds^2 = -\left(1 - \frac{2M}{r} + \frac{Q^2}{r^2} \right) dt^2 + \left(1 - \frac{2M}{r} + \frac{Q^2}{r^2} \right)^{-1} dr^2 + r^2 d\Omega_2^2. \tag{2.6}$$

Comparing (2.2) and (2.6), it can be seen that the metric components g_{tt} and g_{rr} are modified by the additional Q^2/r^2 term. The constant $M > 0$ is once again the mass of the black hole. There is a curvature singularity at $r = 0$.

For $M > |Q|$, there are now two values of r for which the metric component g_{tt} vanishes:

$$r_H = M + \sqrt{M^2 - Q^2}, \qquad r_- = M - \sqrt{M^2 - Q^2}. \tag{2.7}$$

The larger of these two roots, r_H, is the location of the event horizon of the black hole. The smaller root, r_-, is a new type of horizon, known as either an *inner horizon* or a *Cauchy horizon*. If $M = |Q|$, the inner and event horizons merge and an *extremal* black hole results. For $M < |Q|$, there is no event horizon and we have a *naked singularity* at $r = 0$. Penrose diagrams for all these cases can be found in Hawking and Ellis (1975). We will only consider non-extremal black holes for the remainder of this book.

2.2.3 Kerr Black Hole

The two black hole metrics which we have considered thus far have both been static and spherically symmetric. The *Kerr(-Newman)* solution represents a black hole which is rotating. The metric takes the following form, in spheroidal polar co-ordinates (r, θ, φ):

$$ds^2 = -\frac{\Delta}{\Sigma} \left[dt - a \sin^2 \theta \, d\varphi \right]^2 + \frac{\Sigma}{\Delta} dr^2 + \Sigma \, d\theta^2 + \frac{\sin^2 \theta}{\Sigma} \left[\left(r^2 + a^2 \right) d\varphi - a \, dt \right]^2, \tag{2.8}$$

where

$$\Delta = r^2 - 2Mr + a^2 + Q^2, \qquad \Sigma = r^2 + a^2 \cos^2 \theta. \tag{2.9}$$

The constant $M > 0$ is the mass of the black hole. The black hole is rotating, and has angular momentum $J = aM$, where $a > 0$ is a constant. The angle $\theta \in [0, \pi]$ corresponds to latitude: $\theta = 0$ is the rotation axis of the black hole and $\theta = \pi/2$ is the equatorial plane. The metric (2.8) does not depend on time t or the azimuthal (longitude) angle $\varphi \in [0, 2\pi]$. The metric is *stationary*; it is not static because $g_{t\varphi} \neq 0$.

The *Kerr* metric (Kerr 1963) is obtained by setting the charge $Q = 0$, and when $Q \neq 0$ we have the *Kerr-Newman* metric (Newman et al. 1965; Newman and Janis 1965). The Kerr metric is a solution of Einstein's equations (2.1) in a vacuum with $T_{\mu\nu} = 0$. The Kerr-Newman metric is a solution of Einstein's equations with an electromagnetic field, so the stress-energy tensor $T_{\mu\nu}$ has the form (2.4). The electromagnetic field is most compactly written in terms of an electromagnetic potential A_μ, which has non-zero components

$$A_t = -\frac{Qr}{\Sigma}, \qquad A_\varphi = \frac{Qar}{\Sigma}\sin^2\theta. \qquad (2.10)$$

The components of the field strength $F_{\mu\nu}$ can be computed from the electromagnetic potential

$$F_{\mu\nu} = \partial_\mu A_\nu - \partial_\nu A_\mu. \qquad (2.11)$$

As well as an electric part A_t (as in the Reissner-Nordström solution), the electromagnetic potential also has a non-zero magnetic part A_φ, due to the rotation of the black hole.

The metric (2.8) has a curvature singularity when $\Sigma = 0$, that is, $r = 0$ and $\theta = \pi/2$. This is known as the *ring singularity* and its structure is explored in more detail in Hawking and Ellis (1975), O'Neill (1992). The metric component g_{rr} becomes singular when $\Delta = 0$. The values of r at which this happens are, for $M^2 > a^2 + Q^2$,

$$r_H = M + \sqrt{M^2 - a^2 - Q^2}, \qquad r_- = M - \sqrt{M^2 - a^2 - Q^2}. \qquad (2.12)$$

As with the Reissner-Nordström metric, the surface $r = r_H$ is the event horizon of the black hole and $r = r_-$ corresponds to the inner (or Cauchy) horizon. When $M^2 = a^2 + Q^2$, the event and inner horizons merge and the black hole is extremal. For $M^2 < a^2 + Q^2$, there is no event horizon and a naked singularity results.

The event horizon rotates with an angular speed

$$\Omega_H = \frac{a}{r_H^2 + a^2}. \qquad (2.13)$$

The metric component g_{tt} vanishes on a surface $r = r_S$, where

$$r_S = M + \sqrt{M^2 - a^2\cos^2\theta - Q^2}, \qquad (2.14)$$

which lies outside the event horizon $r = r_H$ (see Fig. 2.3). This surface is known as the *stationary limit surface*. The region between the event horizon and the stationary limit surface is known as the *ergosphere*. Within the ergosphere, it is not possible for

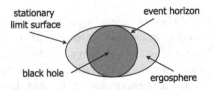

Fig. 2.3 Diagram showing the location of the stationary limit surface and ergosphere for a Kerr black hole. The stationary limit surface lies outside the event horizon of the black hole except on the axis of rotation, where the stationary limit surface touches the event horizon at its north and south poles

a particle to remain at fixed (r, θ, φ). Instead, a massive particle must rotate in the same direction as the event horizon of the black hole. However, since the ergosphere lies outside the event horizon of the black hole, it is still possible for a particle inside the ergosphere to escape from the black hole out to infinity.

The rotation of the black hole described by the Kerr metric and the existence of an ergosphere have many important physical implications. An example is the *Penrose process* (Penrose and Floyd 1971), whereby the rotational energy of the black hole can be extracted. Suppose a particle inside the ergosphere splits into two particles. One of these particles, particle A, escapes to infinity, while the other, particle B, falls down the event horizon and enters the black hole. Imagine that particle B, which falls down the event horizon, has negative energy and angular momentum. Particle B having negative energy means that the energy required to move the particle from its location inside the ergosphere to infinity is greater than the rest-mass energy of the particle. Particle B having negative angular momentum simply means that it is rotating in the opposite direction to the black hole. If particle B has negative energy and angular momentum, particle A, which escapes to infinity, will have greater energy than the original particle. Since particle B enters the event horizon, the energy of the black hole and the angular momentum of the black hole will decrease. The rotational energy of the black hole has effectively been extracted and given to particle A.

Astrophysical black holes are expected to be rapidly rotating, and hence described by the Kerr metric (see, for example, Wiltshire et al. (2009) for more on Kerr black holes in astrophysics). We will also see in Chap. 5 that black holes formed by particle collisions will in general be rapidly rotating.

2.3 Higher-dimensional Black Holes

In Sect. 2.2 we have outlined some of the features of four-dimensional black holes in general relativity. Black holes in more than four space-time dimensions are complex objects in general, as many of the standard theorems governing the properties of four-dimensional black holes do not generalize to higher dimensions. In this section we will discuss the higher-dimensional generalizations of the four-dimensional Schwarz-schild, Reissner-Nordström and Kerr black holes. For more information on the state-of-the-art in research on higher dimensional black holes, including such exotic objects as black rings and black Saturns, see the reviews by Emparan and Reall (2008) and Horowitz (2012). In this section we take the number of space-time dimensions to be $d = n + 4$, so that n represents the number of extra dimensions.

2.3.1 Spherically Symmetric Higher-dimensional Black Holes

The four-dimensional Schwarzschild metric (2.2) has a simple generalization to higher dimensions (Tangherlini 1963):

$$ds^2 = -\left[1 - \left(\frac{r_H}{r}\right)^{n+1}\right]dt^2 + \left[1 - \left(\frac{r_H}{r}\right)^{n+1}\right]^{-1}dr^2 + r^2\,d\Omega_{n+2}^2, \quad (2.15)$$

where $d\Omega_{n+2}^2$ is the metric on the $(n+2)$-sphere. The Schwarzschild-Tangherlini black hole shares many properties with the Schwarzschid black hole. There is an event horizon at $r = r_H$ and a curvature singularity at $r = 0$. The metric (2.15) is static, spherically symmetric and asymptotically flat. The mass M of the black hole is related to the event horizon radius r_H by (Myers and Perry 1986)

$$M = \frac{1}{16\pi}(n+2)\,r_H^{n+1}\,A_{n+2}, \quad (2.16)$$

where A_{n+2} is the area of a unit $(n+2)$-sphere:

$$A_{n+2} = \frac{2\pi^{\frac{n+3}{2}}}{\Gamma\left(\frac{n+3}{2}\right)} \quad (2.17)$$

and $\Gamma(x)$ is the Γ-function. The Reissner-Nordström metric (2.6) also has a straight-forward generalization to higher dimensions (Myers and Perry 1986) but we shall not consider that black hole further.

2.3.2 Higher-dimensional Rotating Black Holes

The generalizations of the neutral Kerr metric (2.8) to higher dimensions are known as *Myers-Perry* black holes (Myers and Perry 1986). The metrics are rather complicated, because a black hole in $(n+4)$ space-time dimensions has N_r independent axes of rotation, where $N_r = k + 1$ if $n = 2k$ is even, and $N_r = k + 2$ if $n = 2k + 1$ is odd. Therefore a four-dimensional Kerr black hole ($n = 0$) has one possible axis of rotation (given by $\theta = 0$ in the Kerr metric (2.8)) but a seven-dimensional black hole has three perpendicular axes of rotation. The metric in the general case can be found in Myers and Perry (1986).

In this book we are interested in black holes formed by a collision of two particles which are moving in a four-dimensional subspace of the higher-dimensional space-time (see Chap. 5). In this situation, if the particles do not collide exactly head-on, the system consisting of the two particles will have non-zero angular momentum about an axis of rotation in the four-dimensional subspace. Therefore, by conservation of angular momentum, the resulting black hole will also have non-zero angular momentum about a single axis, which is in the four-dimensional subspace in which the particles collide. Such black holes are known as *singly rotating* black holes and in this case the general Myers-Perry metric simplifies.

The metric for a singly-rotating Myers-Perry black hole takes the form (Myers and Perry 1986)

$$ds^2 = -\left(1 - \frac{\mu}{\Sigma r^{n-1}}\right) dt^2 - \frac{2a\mu \sin^2\theta}{\Sigma r^{n-1}} dt\, d\varphi + \frac{\Sigma}{\Delta} dr^2 + \Sigma\, d\theta^2$$

$$+ \left(r^2 + a^2 + \frac{a^2\mu \sin^2\theta}{\Sigma r^{n-1}}\right) \sin^2\theta\, d\varphi^2 + r^2 \cos^2\theta\, d\Omega_n^2, \qquad (2.18)$$

where

$$\Delta = r^2 + a^2 - \frac{\mu}{r^{n-1}}, \qquad \Sigma = r^2 + a^2 \cos^2\theta, \qquad (2.19)$$

and $d\Omega_n^2$ is the metric on the n-sphere. The mass M and angular momentum J of the black hole are given by:

$$M = \frac{1}{16\pi}(n+2)\, A_{n+2}\mu, \qquad J = \frac{2}{n+2} aM, \qquad (2.20)$$

so that the parameter $\mu > 0$ governs the mass of the black hole and the parameter $a > 0$ its angular momentum.

The metric component g_{rr} diverges at the event horizon, $r = r_H$, where $\Delta = 0$. For $n > 0$, the function Δ has only one positive real root and therefore the black hole has only an event horizon and no inner horizon. For $n = 1$, the event horizon exists only if $a^2 < \mu$, otherwise there is a naked singularity. For $n > 1$, the event horizon exists for all values of a, so that there is no upper bound on the angular momentum of the black hole.

Higher-dimensional generalizations of the Kerr-Newman metric in closed form have to date eluded researchers. For general relativity with an electromagnetic field, solutions are known only numerically. Analytic solutions are known for higher-dimensional rotating black holes in low-energy models arising from string theory, but these have additional matter fields. See Allahverdizadeh et al. (2011) for a summary of what is known about higher-dimensional charged rotating black holes.

2.4 Brane Black Holes

We now turn to a particular class of higher-dimensional black holes, arising in brane worlds. Firstly we give a very brief summary of the key features of brane worlds before discussing a simple model of the geometry of brane black holes.

2.4.1 Brane Worlds

Brane worlds are a particular class of higher-dimensional theory. There are two main realizations of the brane world idea: the ADD model (Antoniadis et al. 1998; Arkani-Hamed et al. 1998, 1999) in which the extra dimensions are flat, and the

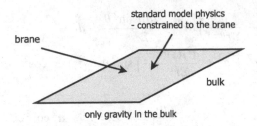

Fig. 2.4 Sketch of a brane world model. The particles and forces of standard model physics (that is, the electromagnetic and strong and weak nuclear forces) are constrained to the brane, only gravity may propagate in the bulk

Randall-Sundrum model (Mannheim 2005; Randall and Sundrum 1999, Randall and Sundrum 1999) in which the extra dimensions are warped (have curvature). These models were proposed as a solution to the hierarchy problem, which can be roughly stated as: why is the force of gravity so much weaker than the other fundamental forces? In brane world models the force of gravity is diluted because it probes the extra dimensions. Quantum gravity in brane world models is discussed in more detail in Chap. 6, here we focus on the nature of classical black holes in these models.

In this section we shall consider only black holes in ADD brane worlds where the extra dimensions have a flat geometry. Matter fields are constrained to be on a four-dimensional subspace (termed the *brane*) of a higher-dimensional space-time (called the *bulk*). Only gravitational degrees of freedom may propagate in the bulk (see Fig. 2.4 for a sketch).

2.4.2 Neutral Brane Black Holes

In the ADD brane world model, the brane can be thought of simply as a "slice" through the higher-dimensional bulk space-time. We shall model the brane itself as having no structure or tension. Now suppose that there is a higher-dimensional black hole in the bulk, as sketched in Fig. 2.5. Since gravity can propagate in the bulk, the higher-dimensional black hole is a solution of Einstein's equations (2.1) in the bulk with $T_{\mu\nu} = 0$ on the right-hand-side (since there is no matter in the bulk, only gravity). The fact that the extra dimensions in the bulk are flat in the ADD model means that the black hole must be asymptotically flat. Assuming that the black hole is much smaller than the size of the extra dimensions, the metric of the higher-dimensional black hole is therefore described by the Myers-Perry metric (see Sect. 2.3.2). If we are interested in higher-dimensional black holes formed by collisions of particles on the brane, the black hole will have a single non-zero component of angular momentum, which will be parallel to the axis of rotation of the black hole in the brane. Therefore the appropriate metric to consider is the singly-rotating Myers-Perry black hole metric (2.18).

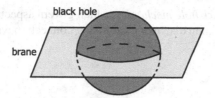

Fig. 2.5 Sketch of a higher-dimensional black hole in an ADD brane world with flat extra dimensions. On taking a slice through the higher-dimensional black hole, the geometry on the brane - a brane black hole - is revealed

To find the metric describing the black hole on the brane, we simply take a slice through the Myers-Perry metric (2.18), by fixing the values of the co-ordinates describing the extra dimensions. The $d\Omega_n^2$ term in the metric (2.18) corresponds to the extra dimensions and disappears when taking this slice, to give the metric of the brane black hole as:

$$ds^2 = -\left(1 - \frac{\mu}{\Sigma r^{n-1}}\right) dt^2 - \frac{2a\mu \sin^2\theta}{\Sigma r^{n-1}} dt\, d\varphi + \frac{\Sigma}{\Delta} dr^2 + \Sigma\, d\theta^2$$
$$+ \left(r^2 + a^2 + \frac{a^2\mu \sin^2\theta}{\Sigma r^{n-1}}\right) \sin^2\theta\, d\varphi^2. \qquad (2.21)$$

If the higher-dimensional black hole is not rotating, then its metric will have the Schwarzschild-Tangherlini form (2.15). Taking a slice through the metric (2.15) gives:

$$ds^2 = -\left[1 - \left(\frac{r_H}{r}\right)^{n+1}\right] dt^2 + \left[1 - \left(\frac{r_H}{r}\right)^{n+1}\right]^{-1} dr^2 + r^2\, d\Omega_2^2. \qquad (2.22)$$

It is worth commenting that the metrics (2.21–2.22) are not solutions of the vacuum Einstein equations (2.1) in four space-time dimensions. Instead, if one calculates the Einstein tensor $G_{\mu\nu}$ for the metrics (2.21–2.22), there is a non-zero stress-energy tensor $T_{\mu\nu}^0$ on the right-hand-side of the Einstein equations (2.1) (Sampaio 2009). This represents an effective fluid seen by an observer on the brane due to the fact that the black hole is actually a higher-dimensional object, but the observer cannot directly probe the extra dimensions.

We use the metrics (2.21–2.22) as simple models for the geometry of brane black holes, particularly in Chap. 3.

2.5 Black Hole Mechanics and Thermodynamics

So far in this chapter we have studied stationary black hole solutions of Einstein's equations. We have not considered black holes as dynamical objects, changing in time. In Chap. 3 we will be interested in evolving black holes. In this section we outline four laws of *black hole mechanics*, which govern aspects of how black holes evolve in time. We will then discover a surprising connection with thermodynamics.

2.5.1 Definitions

Before we can state the four laws of black hole mechanics, a few definitions are needed. Since the defining characteristic of a black hole is that it has an event horizon, the laws of black hole mechanics are essentially laws about the properties of event horizons. We now define two key quantities which can be ascribed to an event horizon.

The first key quantity is the *area* $\mathscr{A}_H$ of an event horizon. This is simply the area of the surface which forms the black hole event horizon. For a four-dimensional Kerr black hole with metric (2.8) the area of the event horizon is

$$\mathscr{A}_H = 4\pi \left(r_H^2 + a^2 \right), \tag{2.23}$$

which reduces to

$$\mathscr{A}_H = 4\pi r_H^2 \tag{2.24}$$

if the black hole is non-rotating (so that $a = 0$).

The next quantity we require is a measure of the acceleration due to gravity at the event horizon of the black hole, known as the *surface gravity* and denoted κ. We are used to the concept of the Newtonian acceleration due to gravity at the surface of a star or planet. In order to consistently define the corresponding quantity in relativity, we have to specify the observer (or frame) in which the acceleration is measured. It would be natural to define the surface gravity as the acceleration of a freely-falling object, instantaneously at rest, near the surface of the body, as measured by an observer at rest near the surface of the body. For a black hole this quantity is divergent at the event horizon, due to the fact that an observer at rest near the event horizon will have a divergent acceleration themselves. However, the acceleration at the event horizon as seen by an observer at infinity is finite, and this will be taken to be the definition of the surface gravity of a black hole. We note that this definition reduces to the usual Newtonian definition in the limit of small velocities and accelerations.

This definition can be used to find the surface gravity of the black holes we have considered earlier in this chapter: see, for example, the treatment in Raine and Thomas (2005). For a four-dimensional Kerr black hole with metric (2.8) the surface gravity is

$$\kappa = \frac{r_H - M}{2Mr_H}. \tag{2.25}$$

For a four-dimensional Schwarzschild black hole with $r_H = 2M$, this reduces to

$$\kappa = \frac{1}{4M}. \tag{2.26}$$

The surface gravity of the higher-dimensional Myers-Perry black holes discussed in Sect. 2.3.2 will also be needed in Chap. 3, and is given by (Myers and Perry 1986):

$$\kappa = \frac{(n+1)\, r_H^2 + (n-1)\, a^2}{2 r_H \left(r_H^2 + a^2\right)}, \tag{2.27}$$

which reduces to (2.25) when $n = 0$.

Finally in this section, we define the *electrostatic potential* Φ_H at the event horizon. For a non-rotating black hole, Φ_H is simply equal to $-A_t$, the time component of the electromagnetic potential. For a rotating black hole,

$$\Phi_H = -A_t - \Omega_H A_\varphi, \tag{2.28}$$

where Ω_H is the angular speed of the event horizon. For a Kerr black hole or charged rotating brane black hole, both of which have the electromagnetic potential (2.10), the electrostatic potential on the horizon is

$$\Phi_H = \frac{Q r_H}{r_H^2 + a^2}, \tag{2.29}$$

which reduces to $\Phi_H = \frac{Q}{r_H}$ when the black hole is non-rotating.

2.5.2 Black Hole Mechanics

With the quantities defined in the previous subsection, we are now in a position to state the laws of black hole mechanics. For simplicity, we restrict attention to four-dimensional black hole solutions of Einstein's equations either in a vacuum or with an electromagnetic field. The four laws of black hole mechanics (Bardeen et al. 1973) are

Zeroth law The surface gravity κ of the event horizon of a stationary black hole is constant over the event horizon.

First law Any two neighbouring stationary, axisymmetric, black hole solutions are related by

$$\delta M = \frac{\kappa}{8\pi} \delta \mathscr{A}_H + \Omega_H \delta J + \Phi_H \delta Q. \tag{2.30}$$

Second law The area $\mathscr{A}_H$ of the event horizon of a black hole is a non-decreasing function of time. If two or more black holes coalesce, the area of the final event horizon is greater than the sum of the areas of the initial event horizons.

Third law It is not possible to form a black hole with a vanishing surface gravity κ in a finite number of operations.

It is clear from (2.25–2.26) that the zeroth law holds for Schwarzschild and Kerr black holes. The zeroth law has been proved in Bardeen et al. (1973) for a class of matter fields which includes electromagnetism (alternative proofs can be found in

DeWitt and DeWitt (1973)). The first law is proved in Bardeen et al. (1973), again for a range of matter fields which includes electromagnetism. The second law is known as the *area theorem* and was proved by Hawking (1972), with some assumptions on the matter fields which are satisfied by electromagnetism. The status of the third law is rather different; it is a postulate and there is currently no mathematical proof, and, equally, no contradiction is known.

Classical processes involving black holes must satisfy the first and second laws of black hole mechanics. For example, in the Penrose process (see Sect. 2.2.3) the event horizon area $\mathscr{A}_H$ does not decrease—see for example the discussion in Raine and Thomas (2005).

2.5.3 Black Hole Entropy and Thermodynamics

The naming and formulation of the laws of black hole mechanics in the previous subsection is rather suggestive of the laws of thermodynamics. In this subsection we explore this analogy further, although it is only in Chap. 3 that the comparison will come to fruition.

Firstly, we observe that classical black holes in general relativity must have an *entropy*. To see why, suppose that black holes did not have an entropy. In this case it would be possible to violate the second law of thermodynamics (that the entropy of the universe cannot decrease) by throwing some entropic matter into a black hole. Therefore black holes must have an entropy and the *generalized second law of thermodynamics* holds, which states that the total entropy of the universe does not decrease, where the total entropy of the universe is the entropy outside black hole event horizons plus the entropy of all the black holes in the universe.

The second law of black hole mechanics suggests that a quantity proportional to the area of the event horizon of a black hole plays the role of black hole entropy (Bardeen et al. 1973; Bekenstein 1973). Let us therefore, in general relativity, set the black hole entropy S_{BH} to be

$$S_{BH} = \mathscr{K}\,\mathscr{A}_H, \tag{2.31}$$

where $\mathscr{K}$ is a constant, independent of the nature of the black hole under consideration.

With this in mind, we now turn to the first law of black hole mechanics (2.30). In relativity mass M and energy E are equivalent (we are using units in which $c = 1$). The second and third terms on the right-hand-side of (2.30) represent the work done in either increasing the angular momentum of the black hole by an amount δJ or increasing the charge of the black hole by an amount δQ. Therefore, the first law of black hole mechanics (2.30) takes the form of the first law of thermodynamics:

$$\delta E = T\delta S + \delta W \tag{2.32}$$

where T is temperature, S is entropy and W is work. If the black hole has an entropy S_{BH} given by (2.31), then it appears that the quantity

$$T_{BH} = \frac{\kappa}{8\pi \mathscr{K}} \qquad (2.33)$$

plays the role of *temperature* in the first law.

For the moment, let us take the approach that T_{BH} is a quantity which is analogous to temperature as far as the laws of black hole mechanics are concerned. The zeroth and third laws of black hole mechanics then immediately have the form of the first and third laws of thermodynamics respectively. The zeroth law implies that a stationary black hole is an equilibrium configuration with constant T_{BH} (corresponding to thermal equilibrium). The third law implies that a "zero temperature" configuration with $T_{BH} = 0$ cannot be reached in a finite number of steps.

2.6 Conclusions

In this chapter we have briefly reviewed the metrics describing four and higher-dimensional black holes in general relativity. We have considered static, spherically symmetric black holes and axisymmetric, rotating black holes. In four-dimensional classical general relativity, black holes are very simple objects, with Kerr-Newman black holes described by their mass, angular momentum and charge. This means that nearly all the information about what has disappeared down the event horizon is lost.

The chapter closed with a brief discussion of the laws of black hole mechanics, which have some similarities with the laws of thermodynamics. This analogy is rather intriguing, but, in general relativity, it is only a metaphor. The temperature of a black hole in general relativity is zero because the black hole, by definition, absorbs matter but emits nothing. We leave the tantalizing question of whether this analogy can be realized as something deeper to Chap. 3.

References

Allahverdizadeh, M., Kunz, J., Navarro-Lerida, F.: J. Phys. Conf. Ser. **314**, 012109 (2011)
Antoniadis, I., Arkani-Hamed, N., Dimopoulos, S., Dvali, G.R.: Phys. Lett. B **436**, 257–263 (1998)
Arkani-Hamed, N., Dimopoulos, S., Dvali, G.R.: Phys. Lett. B **429**, 263–272 (1998)
Arkani-Hamed, N., Dimopoulos, S., Dvali, G.R.: Phys. Rev. D **59**, 086004 (1999)
Bardeen, J.M., Carter, B., Hawking, S.W.: Commun. Math. Phys. **31**, 161–170 (1973)
Bekenstein, J.D.: Phys. Rev. D **7**, 2333–2346 (1973)
Carroll, S.M.: Space-Time and Geometry: An Introduction to General Relativity. Addison Wesley, San Francisco (2003)
DeWitt, C., DeWitt, B.S. (eds.): Black Holes. Taylor and Francis, New York (1973)
Emparan, R., Reall, H.S.: Living Rev. Rel. **11**, 6 (2008)

Frolov, V.P., Novikov, I.D.: Black Hole Physics: Basic Concepts and New Developments. Springer, Berlin (1998)

Frolov, V.P., Zelnikov, A.: Introduction to Black Hole Physics. Oxford University Press, Oxford (2011)

Hartle, J.B.: Gravity: An Introduction to Einstein's General Relativity. Addison Wesley, San Francisco (2002)

Hawking, S.W., Ellis, G.F.R.: The Large-Scale Structure of Space-Time. Cambridge University Press, Cambridge (1975)

Hawking, S.W.: Commun. Math. Phys. **25**, 152–166 (1972)

Hobson, M.P., Efstathiou, G.P., Lasenby, A.N.: General Relativity: An Introduction for Physicists. Cambridge University Press, Cambridge (2006)

Horowitz, G.T (ed.), Black Holes in Higher Dimensions. Cambridge University Press, Cambridge (2012)

Kerr, R.P.: Phys. Rev. Lett. **11**, 237–238 (1963)

Mannheim, P.D.: Brane-Localized Gravity. World Scientific, Singapore (2005)

Misner, C.W., Thorne, K.S., Wheeler, J.A.: Gravitation. W.H. Freeman, San Francisco (1973)

Myers, R.C., Perry, M.J.: Ann. Phys. **172**, 304–347 (1986)

Newman, E.T., Couch, R., Chinnapared, K., Exton, A., Prakash, A., Torrence, R.: J. Math. Phys. **6**, 918–919 (1965)

Newman, E.T., Janis, A.I.: J. Math. Phys. **6**, 915–917 (1965)

Nordström, G.: Verhandl. Koninkl. Ned. Akad. Wetenschap. Afdel. Natuurk. **26**, 1201–1208 (1918)

O'Neill, B.: The Geometry of Kerr Black Holes. A. K. Peters, Wellesley (1992)

Penrose, R., Floyd, R.M.: Nature **229**, 177–179 (1971)

Raine, D.J., Thomas, E.: Black Holes: An Introduction. Imperial College Press, London (2005)

Randall, L., Sundrum, R.: Phys. Rev. Lett. **83**, 3370 3373 (1999)

Randall, L., Sundrum, R.: Phys. Rev. Lett. **83**, 4690–4693 (1999)

Reissner, H.: Annalen der Physik **50**, 106–120 (1916)

Sampaio, M.O.P.: JHEP **0910**, 008 (2009)

Schwarzschild, K.: Sitzungsber. Preuss. Akad. Wiss. Berlin (Math. Phys.), pp. 189–196 (1916).

Tangherlini, F.R.: Nuovo Cim. **27**, 636–651 (1963)

Townsend, P.K.: Black holes, lecture notes from the University of Cambridge. http://arxiv.org/abs/gr-qc/9707012 (1997)

Wald, R.M.: General Relativity. University of Chicago Press, Chicago (1984)

Wiltshire, D.L., Visser, M., Scott, S.M (eds.): The Kerr Space-Time: Rotating Black Holes in General Relativity. Cambridge University Press, Cambridge (2009)

Chapter 3
Hawking Radiation and Black Hole Evaporation

Abstract This chapter is devoted to the most important property of black holes when quantum effects are included, Hawking radiation. Black holes emit quantum radiation with an almost perfect black body spectrum, and therefore ultimately evaporate. We outline the computation and properties of Hawking radiation from both four-dimensional and higher-dimensional black holes.

3.1 Introduction

Classical black holes as described by general relativity (see Chap. 2) are, as the name implies, black. In other words, light and particles can enter the region inside the event horizon but may not escape from inside the event horizon to outside. In quantum mechanics particles are able to propagate into classically forbidden regions. In this chapter we investigate whether quantum particles can escape from a black hole. We treat the black hole space-time as fixed and classical, described by one of the metrics from Chap. 2, and we study a quantum field propagating on this geometry. This is known as the *semi-classical* approximation to quantum gravity, or, alternatively, as *quantum field theory in curved space-time*. Standard monographs on this subject are: Birrell and Davies (1984); Fulling (1989); Mukhanov and Winitzki (2007); Parker and Toms (2009); Wald (1994). The validity of this approximation will be discussed in Sect. 3.2.4.

3.2 Hawking Radiation

Consider a black hole formed by gravitational collapse (see Chap. 4 and 5 for the processes by which black holes may be formed). The details of the collapse process are not important. The Penrose diagram for the space-time is shown in Fig. 3.1

X. Calmet et al., *Quantum Black Holes*, SpringerBriefs in Physics,
DOI: 10.1007/978-3-642-38939-9_3, © The Author(s) 2014

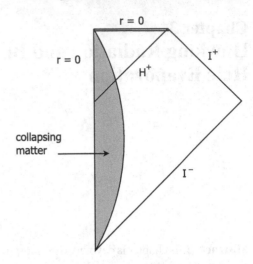

Fig. 3.1 Penrose diagram for a black hole formed by gravitational collapse. The collapsing matter is denoted by the shaded region, and the singularity by the double line at $r = 0$ at the *top* of the diagram. The line $r = 0$ on the *left*-hand-side of the diagram is the centre of the collapsing body. The event horizon is denoted H^+; future null infinity by I^+ and past null infinity by I^-

(compare the Penrose diagram for Schwarzschild space-time in Fig. 2.2). Consider a quantum field on this background geometry. The state of the quantum field is chosen to be the state which is empty of particles in the distant past, before the gravitational collapse has started (this is near past null infinity I^- in Fig. 3.1). Since the space-time is dynamical, this state does not remain empty of particles for all times. Hawking (1975) (see also Jacobson (2003) for an alternative derivation) found that, at late times, far from the black hole (near future null infinity I^+ in Fig. 3.1), a static observer sees an outgoing flux of quantum particles even though initially there are no particles. Effectively, the changing gravitational field due to the formation of the black hole has created quantum particles.

Hawking radiation may be understood heuristically by considering the creation by quantum vacuum fluctuations of a particle-antiparticle pair near the event horizon of the black hole. One of the pair falls down the event horizon; the other escapes to infinity. The particle at infinity must have positive energy as seen by an observer at infinity, so the particle which falls down the horizon must have negative energy as seen by an observer far from the black hole. The observer at infinity therefore sees quantum emission from the black hole. Moreover, this quantum emission seen by an observer far from the black hole has a very special form: that of an (almost) perfect black body.

3.2.1 Black Hole Temperature

Hawking's remarkable result is that black holes have a temperature, known as the *Hawking temperature*. The Hawking temperature for a stationary black hole is (Hawking 1975)

$$T_{BH} = \frac{\kappa}{2\pi}, \tag{3.1}$$

where κ is the surface gravity of the black hole, introduced in Sect. 2.5.1. Using the expressions for the surface gravities (2.25–2.27) of the black holes studied in Chap. 2, the Hawking temperature for all those black holes can be compactly written as

$$T_{BH} = \frac{(n+1)\,r_H^2 + (n-1)\,a^2}{4\pi r_H \left(r_H^2 + a^2\right)}, \tag{3.2}$$

where r_H is the radius of the event horizon of the black hole, a is the angular momentum parameter and n is the number of extra dimensions.

For a Schwarzschild black hole (2.2) the Hawking temperature of the black hole is inversely proportional to its mass M:

$$T_{BH} = \frac{1}{8\pi M}. \tag{3.3}$$

This implies that a black hole has a *negative* specific heat: as it emits Hawking radiation, the black hole loses mass/energy, and its temperature *increases*. The radiation process therefore proceeds at an ever-increasing rate, until the black hole evaporates away completely, leading to black hole explosions (Hawking 1974).

3.2.2 Black Hole Thermodynamics Revisited

In Sect. 2.5.2 we outlined the four laws of black hole mechanics and discovered an intriguing analogy between them and the laws of thermodynamics. As part of this analogy, a quantity $\kappa/8\pi\,\mathcal{K}$ (2.33) was seen to play the same role in the laws of black hole mechanics as played by temperature in the laws of thermodynamics. However, in general relativity, because black holes only absorb and do not emit radiation, this analogy is only a metaphor. In the semi-classical approximation to quantum gravity, we have seen that black holes do in fact have a temperature, which makes the above analogy physical. In particular, comparing (2.33) and (3.1), we see that the unknown constant $\mathcal{K}$ is fixed to be equal to $\mathcal{K} = 1/4$. We therefore find that the entropy of the black hole is

$$S_{BH} = \frac{1}{4}\mathcal{A}_H. \tag{3.4}$$

The laws of black hole mechanics now become the laws of black hole thermodynamics. However, the second law of black hole mechanics does not hold when quantum effects are taken into consideration: a black hole loses mass in Hawking radiation and so its event horizon area shrinks. Accordingly, the entropy of the black hole (3.4) also decreases as Hawking radiation is emitted. The generalized second law of thermodynamics does hold in this situation: the entropy of the universe

exterior to the event horizon plus the black hole entropy does not decrease, due to
the entropy of the Hawking radiation.

Understanding the underlying physics of black hole thermodynamics, including
accounting for the microscopic degrees of freedom which result in the entropy (3.4)
is the subject of much current research, see for example Page (2005); Wald (2001).

3.2.3 Hawking Fluxes

Our focus in this chapter is the Hawking radiation itself, in particular the fluxes of
particles, energy and angular momentum which the black hole emits. These fluxes
are important because the emitted particles can be observed, for example, from
a black hole created in particle collisions in an accelerator (see Chap. 6). In this
scenario the black hole itself cannot be observed, and its properties must be inferred
from the measured Hawking emission. In particular, a detailed understanding of the
Hawking fluxes is needed for accurate simulations of black hole events at accelerators
(Dai et al. 2008; Frost et al. 2009). We will study the Hawking fluxes in detail in
Sects. 3.4 and 3.5, so here we restrict ourselves to a few introductory remarks.

Consider the power spectrum (that is, the energy emitted per unit time) for Hawking emission of a single particle species of spin s from a non-rotating black hole in
$4 + n$ space-time dimensions. This is given by (Kanti 2004)

$$\frac{dE}{dt} = \frac{1}{(2\pi)^{n+3}} \int \sigma_n^s(\omega) \frac{\omega}{e^{\frac{\omega}{T_{BH}}} \pm 1} d^{n+3}k, \tag{3.5}$$

where ω the energy of the emitted degree of freedom and k its momentum. In the
Planck factor, a $+$-sign is used for bosons and a $--$sign for fermions. Restricting
attention to massless fields for which $\omega = |k|$, the integral in (3.5) reduces to an
integral simply over ω:

$$\frac{dE}{dt} = \int \frac{1}{2\pi} \sigma_n^s(\omega) \frac{\omega}{e^{\frac{\omega}{T_{BH}}} \pm 1} \frac{\omega^{n+2}}{2^n \pi^{\frac{(n+1)}{2}} (n+1) \Gamma\left(\frac{n+1}{2}\right)} d\omega. \tag{3.6}$$

The quantity σ_n^s is the *absorption cross-section*. For a perfect black body, it is a
constant corresponding to the area of the body. We will see in Sect. 3.4.3 that a black
hole is not a perfect black body and so σ_n^s will depend on the quantum numbers of
the particle, its spin, and the number of extra dimensions.

3.2.4 Validity of the Semi-classical Approximation

The derivation of the Hawking radiation (Hawking 1975) assumes that the space-time depicted in Fig. 3.1 is classical (that is, described by general relativity). The quantum thermal radiation does, however, have energy, and therefore will affect the space-time geometry. This *back-reaction* is governed by the semi-classical Einstein equations

$$G_{\mu\nu} = 8\pi \langle \hat{T}_{\mu\nu} \rangle, \qquad (3.7)$$

where on the right-hand-side we have the expectation value of the quantum stress-energy tensor operator. The back-reaction of the quantum field can be incorporated in a simple way by assuming that the Hawking radiation is a continuous process and that the black hole retains the Schwarzschild form (2.2) as it evolves. If, in a small interval of time δt, the black hole emits energy $\delta\mu$ in Hawking radiation, then its mass decreases by $\delta\mu$. This is a reasonable approximation if $\delta\mu \ll M$, so that the energy of each emitted quantum of radiation is much smaller than the mass of the black hole.

3.3 Stages of Evolution of a Microscopic Black Hole

We now consider a microscopic black hole formed by the high-energy collision of particles (see Chaps. 5 for details of this process), either in a particle accelerator or in cosmic rays. When initially formed, the black hole will be rapidly spinning and highly asymmetric. If, as in a particle accelerator, the black hole forms from colliding particles which have electromagnetic and colour charges, the black hole will initially also have electromagnetic and colour charges, which are known as *gauge field hair*. The subsequent evolution of the black hole is a continuous process, but it is helpful to model the evolution as involving four distinct stages (Giddings and Thomas 2002)—see Fig. 3.2 for sketches of the different stages:

Balding phase During this phase the black hole sheds its gauge field hair by the emission of particles with gauge field charges, and loses its asymmetries by the emission of gravitational radiation. Usually this phase is regarded as part of the formation process of the black hole.

Spin-down phase At the end of the balding phase, the black hole is still rapidly rotating, but it is axisymmetric. The metric of the black hole is described by the higher-dimensional Myers-Perry metric (2.18). During this phase, the black hole emits Hawking radiation and loses its angular momentum. The black hole also shrinks in size and mass.

Schwarzschild phase At the end of the spin-down phase, the black hole has lost all its angular momentum. The space-time is now spherically symmetric, and described by the higher-dimensional Schwarzschild-Tangherlini metric (2.15). The black hole continues to emit Hawking radiation and thereby lose mass.

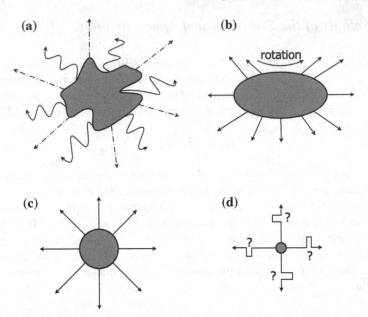

Fig. 3.2 Sketch of the four stages of evolution of a microscopic black hole: (**a**) balding phase, (**b**) spin-down phase, (**c**) Schwarzschild phase and (**d**) Planck phase. In (**a**), the *curly arrows* denote gauge field hair attached to the black hole and the *dashed arrows* denote the emission of gravitational radiation. In (**b**) and (**c**) the *arrows* indicate Hawking radiation from the black hole. In (**d**) the *arrows* and question marks indicate that the behaviour in this phase is not fully understood

Planck phase The Hawking radiation in the spin-down and Schwarzschild phases is treated semi-classically. As discussed in Sect. 3.2.4, this approximation is only valid when the emitted quanta have energy much smaller than the mass of the black hole. Towards the end of the life of the black hole, the black hole will be so light that each emitted quantum will have energy which is a significant fraction of the energy of the black hole and the semi-classical approximation will no longer be valid. In this phase the details of the (as yet unknown) theory of quantum gravity become important and each particle emitted in Hawking radiation will have a dramatic effect on the black hole. The physics of this phase of the evolution of the black hole is therefore poorly understood at present. See Sect. 5.4 for discussion of planckian quantum black holes.

For the remainder of this Chapter, we study in more detail the *spin-down* and *Schwarz-schild* phases of the evolution, where the semi-classical approximation is valid. We will discuss the nature of the Hawking radiation in these two phases.

3.4 Computation of Hawking Radiation

The Hawking radiation emitted by a semi-classical black hole is not precisely thermal. This is because there is an effective gravitational potential surrounding the black hole, through which each quantum particle must pass if it is to travel out to infinity, far from the black hole. This gravitational potential is felt by classical waves as well, and part of a wave outgoing from the event horizon will be reflected back down the event horizon by the gravitational potential, while part is transmitted to infinity.

In order to account for this gravitational scattering, the behaviour of classical waves for each type of quantum field must be studied in detail. We are interested in the emission of scalars (spin $s = 0$), fermions ($s = \frac{1}{2}$), gauge bosons (including photons, $s = 1$) and gravitons (quanta of the gravitational field, $s = 2$).

In this section we outline the formalism used to study these scattering processes, and write down some of the key equations. Fuller treatments can be found in the literature: Chandrasekhar (1998) for classical fields on four-dimensional black holes and Kanti (2004) for fields on the brane. In brane world models (see Sect. 2.4.1), standard model particles (scalars, fermions and gauge bosons) are restricted to the brane, whereas gravitational degrees of freedom (gravitons and possibly scalars) can propagate in the higher-dimensional bulk. Our summary is split into two sections: firstly we discuss the formalism for fields of spin-0, $\frac{1}{2}$ and 1 on the brane, and secondly fields of spin-0 and 2 in the bulk. We will then describe the impact of the scattering processes described above on the Hawking fluxes introduced in Sect. 3.2.3. Since there a number of different quantum fields and different scenarios to consider, finding a consistent notation is a challenge. We will index many quantities by Λ. The exact nature of the index Λ will depend on the particular field under consideration, and will be given in the relevant subsection.

3.4.1 Teukolsky Formalism on the Brane

In brane-world models, particles of spin $s = 0, \frac{1}{2}, 1$ are restricted to move on a four-dimensional brane black hole geometry. In this subsection we outline the formalism for these fields on the brane black hole metric (2.21). The equations in this subsection also apply to four-dimensional Schwarzschild and Kerr black holes, on setting the number of extra dimensions $n = 0$.

For four-dimensional Kerr black holes ($n = 0$, see Sect. 2.2.3), Teukolsky (1972, 1973) has developed a unified formalism for describing perturbations of spin $s = 0, \frac{1}{2}$, 1 and 2 (see also Chandrasekhar (1998) for a comprehensive treatment). The approach involves applying the Newman-Penrose formalism to write the perturbation equations for each spin as a single master equation for a variable $\Psi_s = \Psi_s(t, r, \theta, \varphi)$ which represents the spin-field perturbation, depending on the space-time co-ordinates (t, r, θ, φ). Teukolsky's formalism readily extends to peturbations of spin $s = 0$, $s = \frac{1}{2}$ and $s = 1$ on a brane slice of a higher-dimensional, rotating Myers-Perry black hole (2.21). The details of the derivation of the master equation in this case can be found in Kanti (2004). The precise definitions of the variable Ψ_s depend on the particular quantum field being considered: the forms can be found in Kanti (2004).

Due to remarkable symmetry properties of the four-dimensional metric (2.8, 2.22) the Teukolsky master equation turns out to be separable. Writing

$$\Psi_s = e^{-i\omega t} e^{im\varphi} R_\Lambda(r) S_\Lambda(\theta), \tag{3.8}$$

the following radial and angular equations are obtained (Casals et al. 2007; Kanti 2004):

$$0 = \Delta^{-s} \frac{d}{dr} \left(\Delta^{s+1} \frac{dR_\Lambda}{dr} \right) + \left[\Delta^{-1} \left(K_{\omega m}^2 - is K_{\omega m} \Delta' \right) + 4is\omega r \right.$$

$$+ s\delta_{s,|s|} \left(\Delta'' - 2 \right) - a^2\omega^2 + 2ma\omega - \lambda_\Lambda \Big] R_\Lambda(r), \tag{3.9}$$

$$0 = \frac{1}{\sin\theta} \frac{d}{d\theta} \left(\sin\theta \frac{dS_\Lambda}{d\theta} \right) + \left[-\frac{2ms\cot\theta}{\sin\theta} - \frac{m^2}{\sin^2\theta} + a^2\omega^2\cos^2\theta - 2as\omega\cos\theta \right.$$

$$+ s - s^2\cot^2\theta + \lambda_\Lambda \Big] S_\Lambda(\theta), \tag{3.10}$$

where

$$K_{\omega m} = \left(r^2 + a^2 \right) \omega - am. \tag{3.11}$$

Here, ω is the energy of the particular wave mode and m is the azimuthal quantum number. We define an index $\ell = s, s+1, \ldots$, which labels the angular momentum quantum numbers which index the separation constants λ_Λ. In this case the index Λ on the radial and angular functions and the separation constants is $\Lambda = \{s, \omega, \ell, m\}$, and so depends on the spin of the field s as well the quantum numbers. The function Δ appearing in the radial equation (3.9) comes from the metric (2.9, 2.19).

The angular equation (3.10) is the same as for a four-dimensional Kerr black hole. The angular functions $S_\Lambda(\theta)$ are spin-weighted spheroidal functions. The eigenvalues λ_Λ are complicated functions of $a\omega$ which have to be found numerically by solving (3.10) subject to the boundary conditions of regularity on the axis of rotation of the black hole, where $\theta = 0, \pi$. For $a\omega = 0$, their values are known:

$$\lambda_\Lambda = \ell(\ell+1) - s(s+1), \tag{3.12}$$

and the spin-weighted spheroidal harmonics reduce to spin-weighted spherical harmonics.

3.4.2 Bulk Fields

In brane world models, gravity is the only force which propagates into the higher-dimensional bulk. In this subsection we outline what is known about gravitational perturbations of higher-dimensional rotating Myers-Perry black holes (see Sect. 2.3.2). We also consider scalar field emission in the bulk.

On the full $4 + n$-dimensional Myers-Perry metric (2.18), the scalar ($s = 0$) wave equation can easily be derived and is separable. The scalar field Ψ_0 is written as

$$\Psi_0 = e^{-i\omega t} e^{im\varphi} R_\Lambda(r) S_\Lambda(\theta) Y_{jn} \qquad (3.13)$$

where, as above, (t, r, θ, φ) are the co-ordinates on the brane and Y_{jn} is a hyperspherical harmonic depending on the higher-dimensional bulk co-ordinates and indexed by an integer j. The following radial and angular equations are then obtained (Casals et al. 2008):

$$0 = \frac{1}{r^n} \frac{d}{dr} \left(r^n \Delta \frac{dR_\Lambda}{dr} \right) + \left[\Delta^{-1} K^2_{\omega m} - a^2 r^{-2} j (j + n - 1) \right.$$
$$\left. - a^2 \omega^2 + 2ma\omega - \lambda_\Lambda \right] R_\Lambda(r), \qquad (3.14)$$

$$0 = \frac{1}{\sin\theta \cos^n \theta} \frac{d}{d\theta} \left(\sin\theta \cos^n \theta \frac{dS_\Lambda}{d\theta} \right) + \left[\omega^2 a^2 \cos^2 \theta - \frac{m^2}{\sin^2 \theta} - \frac{j (j + n - 1)}{\cos^2 \theta} \right.$$
$$\left. + \lambda_\Lambda \right] S_\Lambda(\theta), \qquad (3.15)$$

which can be seen to reduce to the brane $s = 0$ radial (3.9) and angular (3.10) equations when the number of extra dimensions $n = 0$. In this case the index $\Lambda = \{\omega, \ell, m, j, n\}$.

The formalism for bulk gravitational perturbations (spin $s = 2$) does not readily generalize from Teukolsky's work. For a spherically symmetric higher-dimensional black hole with metric (2.15), the master equations have been derived by Kodama and Ishibashi (2003). A general gravitational perturbation is decomposed into a symmetric traceless tensor, a vector and a scalar part. For each type of gravitational perturbation, the master equation is separable, writing the relevant field quantity in a form similar to (3.13). The angular functions of each type of gravitational perturbation are spin-weighted spherical harmonics, and the radial functions satisfy the following equation (Ishibashi and Kodama 2003)

$$0 = \left[1 - \left(\frac{r_H}{r} \right)^{n+1} \right] \frac{d}{dr} \left\{ \left[1 - \left(\frac{r_H}{r} \right)^{n+1} \right] \frac{dR_\Lambda}{dr} \right\} + \left[\omega^2 - \mathcal{V}_\Lambda \right] R_\Lambda(r), \quad (3.16)$$

where the form of the potential $\mathcal{V}_\Lambda$ depends on the type of gravitational perturbation. In this case the index $\Lambda = \{B, \omega, \ell, n\}$ where $B \in \{S, V, T\}$ indicates whether we are considering a scalar (S), vector (V) or tensor (T) type of gravitational perturbation.

The potential $\mathcal{V}_\Lambda$ has the form (Ishibashi and Kodama 2003)

$$\mathscr{V}_{T/V,\omega,\ell,n} = \frac{1}{r^2}\left[1-\left(\frac{r_H}{r}\right)^{n+1}\right]\left[\ell\,(\ell+n+1)+\frac{n\,(n+2)}{4}-\frac{k}{4}\,(n+2)^2\,\frac{r_H^{n+1}}{r^{n+1}}\right]$$

$$(3.17)$$

for tensor-like $(T, k = -1)$ and vector-like $(V, k = 3)$ perturbations. For scalar-like (S) graviton perturbations, the potential is more complicated (Ishibashi and Kodama 2003):

$$\mathscr{V}_{S,\omega,\ell,n} = \frac{1}{r^2}\left[1-\left(\frac{r_H}{r}\right)^{n+1}\right]\frac{qx^3+px^2+wx+z}{4\,[2u+(n+2)(n+3)x]^2} \qquad (3.18)$$

where

$$x = \frac{r_H^{n+1}}{r^{n+1}}, \qquad u = \ell\,(\ell+n+1)-n-2, \qquad (3.19)$$

and

$$q = (n+2)^4\,(n+3)^2\,,$$
$$p = (n+2)\,(n+3)\left[4u\left(2n^2+5n+6\right)+n\,(n+2)\,(n+3)\,(n-2)\right],$$
$$w = -12u\,(n+2)\,[u\,(n-2)+n\,(n+2)\,(n+3)]\,,$$
$$z = 16u^3+4u^2\,(n+2)\,(n+4)\,. \qquad (3.20)$$

The radial equation (3.16) has the form of a standard Schrödinger equation. In this case the gravitational potential barrier surrounding the black hole can be clearly seen in the equation.

For rotating higher-dimensional black holes, the general gravitational perturbations are proving elusive—the only ones known for a singly-rotating black hole, whose metric is given in Chap. 2 are those for tensor-type perturbations (Kodama 2008, 2009). In this case, the radial and angular equations for the tensor-type graviton modes are identical to (3.14–3.15) for the scalar field modes - the only difference is that $\ell \geq 0$ for scalars and $\ell \geq 2$ for gravitons (Kodama 2009).

3.4.3 Greybody Factors

In order to compute the Hawking emission of particles, energy and angular momentum, the radial equations described in Sects. 3.4.1 and 3.4.2 above need to be solved numerically for each type of quantum field. The ultimate goal is to find the *greybody factor* (also known as the *absorption probability* or *transmission coefficient*), which is the proportion of a wave propagating outwards from the event horizon which passes through the gravitational potential and reaches infinity.

The greybody factor $\mathscr{T}_\Lambda$ is the ratio of the flux in the out-going wave at infinity $\mathscr{F}_\Lambda^\infty$ and the flux in the out-going wave near the event horizon $\mathscr{F}_\Lambda^H$:

$$\mathscr{T}_\Lambda = \frac{\mathscr{F}_\Lambda^\infty}{\mathscr{F}_\Lambda^H}. \tag{3.21}$$

The greybody factor depends on the appropriate quantities Λ indexing each type of field. The method for computing the flux varies depending on the nature of the field considered (the details can be found in Chandrasekhar (1998); Kanti (2004)) and the results are summarized below. In each case the greybody factor is constructed from an appropriate solution of the relevant radial equation (3.9, 3.14, 3.16).

We consider first scalar fields (both on the brane and in the bulk) and graviton fields. For scalar fields, there is just one radial function R_Λ, satisfying either (3.9) on the brane or (3.14) in the bulk. For graviton fields, each type of perturbation for which the master equation is currently known is described by a single radial function R_Λ, satisfying either (3.16) for all types of graviton emission from a non-rotating black hole, or (3.14) for tensor-type graviton emission from a singly-rotating black hole. In these two cases, we consider a wave solution of the relevant radial equation having the following leading-order behaviour:

$$R_\Lambda \sim \begin{cases} (r - r_H)^{i\tilde\omega/4\pi T_{BH}} + C_{R,\Lambda}\,(r - r_H)^{-i\tilde\omega/4\pi T_{BH}} & r \to r_H \\ C_{T,\Lambda}\,r^{-y}e^{i\omega r} & r \to \infty, \end{cases} \tag{3.22}$$

where $C_{R,\Lambda}$ and $C_{T,\Lambda}$ are complex constants,

$$\tilde\omega = \omega - m\Omega_H \tag{3.23}$$

with Ω_H is the angular velocity of the black hole event horizon, and

$$y = \begin{cases} 1 & \text{for brane emission of scalars,} \\ 1 + \frac{n}{2} & \text{for bulk emission of scalars, and tensor-type graviton emission} \\ & \text{from rotating black holes,} \\ 0 & \text{for graviton emission from non-rotating black holes.} \end{cases} \tag{3.24}$$

Near the event horizon, the energy of the wave is $\tilde\omega$ rather than ω (the energy at infinity) due to the rotation of the black hole. If the black hole is non-rotating, then $\Omega_H = 0$ and $\tilde\omega = \omega$.

The radial function (3.22) represents a wave propagating outwards from the event horizon, part of which (the part containing the constant $C_{R,\Lambda}$) is reflected by the gravitational potential back down the event horizon, and part (the part containing the constant $C_{T,\Lambda}$) is transmitted out to infinity.

For a radial function R_Λ of the form (3.22) the greybody factor is simply

$$\mathscr{T}_\Lambda = 1 - \left| C_{R,\Lambda} \right|^2. \tag{3.25}$$

For scalar and graviton perturbations, from the properties of the radial equations (3.9, 3.14, 3.16) it can be shown that

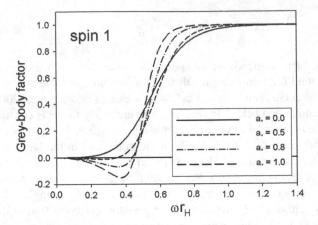

Fig. 3.3 Typical greybody factors for a spin-1 field, as a function of the wave energy ω. We consider a rotating brane black hole with metric (2.21) and $n = 1$. Four different values of the angular speed of the event horizon are considered, with $a_* = a/r_H$. The results are shown for the field mode with $\ell = m = 1$

$$1 - \left| C_{R,\Lambda} \right|^2 = \frac{\omega}{\tilde{\omega}} \left| C_{T,\Lambda} \right|^2 , \qquad (3.26)$$

where $\tilde{\omega}$ is given by (3.23). If the black hole is non-rotating, $\tilde{\omega} = \omega$ and the right-hand-side of (3.26) reduces to $|C_{T,\Lambda}|^2$. When the black hole is rotating, for modes with $\omega/\tilde{\omega} < 0$, Eq. (3.26) implies that $|C_{R,\Lambda}|^2 > 1$ and $\mathscr{T}_\Lambda < 0$. This is the phenomenon of *super-radiance* (Chandrasekhar 1998), whereby a wave incident on a rotating black hole can be reflected back to infinity with an amplitude greater than it had initially. Similarly, a wave propagating outwards from the event horizon can be reflected back down the event horizon with an amplitude greater than it had initially (see Fig. 3.3). Super-radiance is the wave analogue of the Penrose process (see Sect. 2.2.3).

For fermion (spin-$\frac{1}{2}$) and gauge boson (spin-1) fields, the calculation of the greybody factor is more involved. In these cases there are two radial functions, corresponding to $s = +|s|$ and $s = -|s|$. The radial function R_Λ satisfies (3.9) and has the asymptotic forms (Casals et al. 2007; Kanti 2004)

$$R_\Lambda \sim \begin{cases} B_{\mathrm{in}} \Delta^{-s} (r - r_H)^{-i\tilde{\omega}/4\pi T_{BH}} + B_{\mathrm{out}} (r - r_H)^{i\tilde{\omega}/4\pi T_{BH}} & r \to r_H \\ C_{\mathrm{in}} r^{-\delta_{s,1}-2s} e^{-i\omega r} + C_{\mathrm{out}} r^{-\delta_{s,1}} e^{i\omega r} & r \to \infty, \end{cases}$$

$$\qquad (3.27)$$

where $B_{\mathrm{in/out}}$ and $C_{\mathrm{in/out}}$ are complex constants. From (3.27) we see that the two radial functions have different asymptotic forms: for $s = +|s|$, the dominant mode is incoming (containing the constants B/C_{in}) while for $s = -|s|$ the dominant mode is outgoing (containing the constants B/C_{out}). To compute the greybody factor, we consider solutions of the radial equation of the form

$$R_\Lambda^{s=+|s|} \sim \begin{cases} C_{R,\Lambda}\Delta^{-s}(r-r_H)^{-i\tilde{\omega}/4\pi T_{BH}} & r \to r_H \\ 0 & r \to \infty, \end{cases}$$

$$R_\Lambda^{s=-|s|} \sim \begin{cases} (r-r_H)^{i\tilde{\omega}/4\pi T_{BH}} & r \to r_H \\ C_{T,\Lambda}r^{-\delta_{s,1}}e^{i\omega r} & r \to \infty, \end{cases} \tag{3.28}$$

for complex constants $C_{R,\Lambda}$ and $C_{T,\Lambda}$. The greybody factor $\mathscr{T}_\Lambda$ is then given by (3.25). For gauge bosons ($|s| = 1$), the relation (3.26) holds and we have super-radiance as for the scalar and graviton fields. However, for fermion fields ($|s| = \frac{1}{2}$), the relation equivalent to (3.26) takes the form

$$1 - |C_{R,\Lambda}|^2 = |C_{T,\Lambda}|^2. \tag{3.29}$$

Therefore, for fermions it is always the case that $|C_{R,\Lambda}|^2 < 1$ and $\mathscr{T}_\Lambda > 0$. There is no super-radiance for fermion fields (Chandrasekhar 1998).

As an example, in Fig. 3.3 we show some greybody factors for a spin-1 field as a function of the wave energy ω, for a rotating black hole. For low-energy waves, as $\omega \to 0$, the greybody factor tends to zero as the out-going wave is entirely reflected back down the event horizon. For high-energy waves, as $\omega \to \infty$, all the wave is trasmitted through the potential barrier to infinity. When the black hole is rotating ($a_* \neq 0$ in Fig. 3.3), it can be seen that there is a range of values of ω for which the greybody factor is negative. This range of values of ω corresponds to $\tilde{\omega} < 0$ (3.23) and therefore, from (3.25–3.26), the greybody factor is negative and we have super-radiance.

3.4.4 Emission of Particles, Energy and Angular Momentum

Once the greybody factors have been calculated, the fluxes of particles, energy and angular momentum emitted by a black hole in Hawking radiation can be computed. Consider the expression for the power spectrum from a non-rotating black hole (3.6). The absorption cross-section σ_n^s is given in terms of the greybody factors as (Kanti 2004)

$$\sigma_n^s(\omega) = \frac{2^n \pi^{\frac{(n+1)}{2}}(n+1)\,\Gamma\left(\frac{n+1}{2}\right)}{\omega^{n+2}} \sum_{\text{modes}} \mathscr{T}_\Lambda, \tag{3.30}$$

where the sum is taken over all field modes with energy ω as measured from infinity. Substituting (3.30) into (3.6), it can be seen that the power spectrum has a simple form in terms of the greybody factors.

Similar expressions can be derived for the spectra of emitted particles and angular momentum. For the particle spectrum, there is no ω in the numerator of the fraction containing the Planck factor in (3.6), while for the angular momentum spectrum the factor of ω in the numerator of the fraction containing the Planck factor is replaced by m, the azimuthal quantum number of the field mode (3.8, 3.13). We can also extend the formula (3.6) to rotating black holes. Since Hawking radiation emanates from the horizon of a black hole, the thermal factor depends on the energy $\tilde{\omega}$ (3.23) of a field mode near the horizon, which is different from its energy at infinity, ω, if the black hole is rotating.

To summarize, the differential fluxes per unit time and unit energy of particles N, energy E and angular momentum J for the emission of particles of spin s can be written in the following simple forms:

$$\frac{d^2 N}{dt\, d\omega} = \frac{1}{2\pi} \sum_j \sum_{\ell=s}^{\infty} \sum_{m=-\ell}^{\ell} \frac{1}{e^{\tilde{\omega}/T_{BH}} \pm 1} \mathcal{N}_\Lambda \mathcal{T}_\Lambda, \tag{3.31}$$

$$\frac{d^2 E}{dt\, d\omega} = \frac{1}{2\pi} \sum_j \sum_{\ell=s}^{\infty} \sum_{m=-\ell}^{\ell} \frac{\omega}{e^{\tilde{\omega}/T_{BH}} \pm 1} \mathcal{N}_\Lambda \mathcal{T}_\Lambda, \tag{3.32}$$

$$\frac{d^2 J}{dt\, d\omega} = \frac{1}{2\pi} \sum_j \sum_{\ell=s}^{\infty} \sum_{m=-\ell}^{\ell} \frac{m}{e^{\tilde{\omega}/T_{BH}} \pm 1} \mathcal{N}_\Lambda \mathcal{T}_\Lambda. \tag{3.33}$$

The plus sign in the thermal Planck factors in the denominators of the fractions is applicable for fermionic fields and the minus sign for bosonic fields. The mode sums are taken over all values of the azimuthal quantum number m (3.8, 3.13) and the angular momentum quantum number ℓ. The additional sum over j, which indexes the hyperspherical harmonics in the bulk, is absent for fields on four-dimensional black holes. It is present only for scalar fields in the bulk and tensor-like graviton emission from a rotating black hole. For graviton emission from a non-rotating higher-dimensional black hole, the additional dimensions are taken into account in the angular momentum quantum number ℓ of the field mode.

In (3.31–3.33), as well as the greybody factor $\mathcal{T}_\Lambda$, the fluxes also contain a degeneracy factor $\mathcal{N}_\Lambda$ which counts the multiplicity of modes having the quantum numbers $\{\ell\omega m\}$ (for all fields) and j for bulk fields. As previously, the exact form of the label Λ on the degeneracy factor depends on the nature of the field under consideration. For reference, we now list the degeneracy factors for the different types of field. None of the degeneracy factors depend on the mode energy ω, although this is included in the label Λ for consistency with previous notation.

Brane fields. For fields of spin-$\frac{1}{2}$ and spin-1, there are field modes with two polarizations, so to take this into account we set the degeneracy factors equal to

$$\mathcal{N}_\Lambda = \begin{cases} 1 & \text{for } s = 0, \\ 2 & \text{for } s = \frac{1}{2}, \\ 2 & \text{for } s = 1, \end{cases} \tag{3.34}$$

where the above hold for any values of ω, ℓ and m.

Bulk scalar fields: For bulk scalar fields, the degeneracy factor depends on the index j which labels the hyperspherical harmonics (3.13) and counts the multiplicity of modes in the bulk (Casals et al. 2008):

$$\mathcal{N}_\Lambda = \frac{(2j + n - 1)(j + n - 2)!}{j!(n-1)!}. \tag{3.35}$$

Note that the degeneracy factor is the same for all ω, ℓ and m.

Bulk graviton fields: For bulk graviton fields, the number of degrees of freedom increases significantly as the number of space-time dimensions increases. For gravitational perturbations of spherically symmetric black holes, the degeneracy factors $\mathcal{N}_\Lambda$ depend on the type of gravitational perturbation (scalar, vector or tensor-like) and count the number of modes with the same angular momentum quantum number ℓ. Their values are (Creek et al. 2006):

$$\mathcal{N}_{S,\omega,\ell,n} = \frac{(2\ell + n + 1)(\ell + n)!}{(2\ell + 1)\ell!(n+1)!},$$

$$\mathcal{N}_{V,\omega,\ell,n} = \frac{\ell(\ell + n + 1)(2\ell + n + 1)(\ell + n - 1)!}{(2\ell + 1)(\ell + 1)!n!},$$

$$\mathcal{N}_{T,\omega,\ell,n} = \frac{n(n+3)(\ell + n + 2)(\ell - 1)(2\ell + n + 1)(\ell + n - 1)!}{2(2\ell + 1)(\ell + 1)!(n+1)!}. \tag{3.36}$$

For tensor-type gravitational perturbations of rotating black holes, the degeneracy factor is (Kanti et al. 2009)

$$\mathcal{N}_\Lambda = \frac{(n+1)(n-2)(n+j)(j-1)(n+2j-1)(n+j-3)!}{2(j+1)!(n-1)!}. \tag{3.37}$$

As with bulk scalar fields, the degeneracy factor depends on the index j which labels the hyperspherical harmonics in the bulk but does not depend on the quantum numbers ℓ or m.

3.5 Emission of Neutral, Massless Particles

We now describe the physical properties of the fluxes of particles, energy and angular momentum (3.31–3.33) due to Hawking radiation. In this section we focus on the emission of neutral, massless particles, and we will briefly discuss more general emission in the following section. We begin with the emission from a four-dimensional black hole, which was studied in depth by (Page 1976, 1977). Over the past ten years, Hawking radiation from higher-dimensional black holes has been studied in depth and there is a very large literature on this subject. Relevant review articles include those by Casanova and Spallucci (2006); Kanti (2004, 2009, 2012); Landsberg (2004); Winstanley (2007).

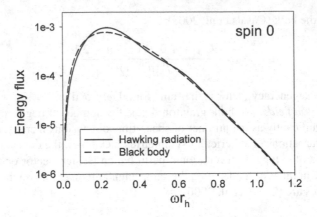

Fig. 3.4 Energy flux (3.32) for the emission of scalar particles from a four-dimensional Schwarz-schild black hole. The *solid line* denotes the flux (3.32) including the greybody factors and the *dashed lines* denote the pure black body spectrum for a black body having the same temperature as the black hole and a constant absorption cross-section $\sigma_0^0 = \frac{27\pi r_h^2}{4}$

3.5.1 Emission from Four-dimensional Black Holes

To illustrate the typical form of the spectrum of Hawking radiation, in Fig. 3.4 we show the energy flux (3.32) for scalar emission from a four-dimensional Schwarz-schild black hole (see Sect. 2.2.1), for which the Hawking temperature is given by (3.3). We plot both the exact Hawking flux, computed using the greybody factors, and a pure black body spectrum at the same temperature. The importance of including the greybody factors can be seen in the low-energy emission. At high energies, since the greybody factors approach unity (see Fig. 3.3), the pure black body spectrum becomes a better approximation. The power spectra (energy fluxes) for the emission of other particles (fermions, gauge bosons and gravitons) from a Schwarzschild black hole have very similar shapes (Page 1976). As the spin of the field increases, the height of the peak of emission decreases and the peak occurs at higher energies (Page 1976). The particle fluxes also have a similar shape. Since a Schwarzschild black hole is non-rotating, no angular momentum flux is emitted in the Hawking radiation.

If we consider the energy flux (3.32) from a rotating Kerr black hole (see Sect. 2.2.3), the shape of the spectrum is very different from that for a non-rotating black hole—see Fig. 3.5. The power spectrum has a number of peaks, which are due to particular modes becoming dominant at different energies ω. The spectra for fields of other spin have similar features, although the magnitude of the largest peak varies according to the spin, as does the rate at which the energy flux dies off for large ω. The temperature of a four-dimensional Kerr black hole is given by (3.2) with $n = 0$, that is

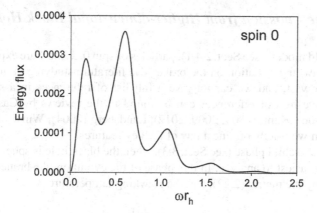

Fig. 3.5 Energy flux (3.32) for the emission of scalar particles from a four-dimensional Kerr black hole. The rotation parameter is $a = 0.5r_h$

$$T_{BH} = \frac{r_H^2 - a^2}{4\pi r_H \left(r_H^2 + a^2\right)}, \tag{3.38}$$

so that the temperature depends on the angular momentum parameter a as well as the mass of the black hole. For fixed radius of the event horizon r_H, the temperature (3.38) is a maximum when $a_* = a/r_H = 0$ (that is, the black hole is not rotating) and decreases as a_* increases, until the temperature vanishes when $a_* = 1$ and the black hole is extremal. This variation in temperature affects the relative emissions of different types of particle. This has been studied in detail by Page (1976).

The fluxes of particles and angular momenta have similar shapes, as do the spectra for particles with non-zero spin. In Page (1976), the emission of massless fermions, gauge bosons (that is, photons) and gravitons from a Kerr black hole is studied in detail. The black hole sheds mass and angular momentum, but sheds angular momentum more quickly than mass. For example, a black hole starting off with very close to the maximal rotation $a = M$, emitting four species of massless fermion, one species of photon and one species of graviton will lose half its mass in 71 % of its lifetime but half its angular momentum in only 6.7 % of its lifetime. Page (1976) finds that the black hole loses angular momentum so quickly that, if it is emitting the combination of species outlined above, it will "emit more than 50 % of its energy when it is so slowly rotating that its power is within 1 % of the Schwarzschild value" Page (1976), page 3267.

This means that, for four-dimensional black holes, the "spin-down" phase (see Sect. 3.3) of the evolution of a black hole is very rapid, and the Schwarzschild phase is dominant.

3.5.2 Brane Emission from Higher-dimensional Black Holes

In brane-world models (see Sect. 2.4.1), particles of spin 0, $\frac{1}{2}$ and 1 are expected to be emitted in Hawking radiation on the brane. The literature studying brane emission is now rather vast, and we cannot give a full list of references in a short book. Comprehensive lists of references can be found in the reviews by Casanova and Spallucci (2006); Kanti (2004, 2009, 2012); Landsberg (2004); Winstanley (2007). In this section we briefly outline a few of the key features.

The Schwarzschild phase (see Sect. 3.3), when the black hole is spherically symmetric, is the easiest to analyze. In this phase of the evolution, the brane black hole is described by the metric (2.22), and the Hawking temperature is (3.2)

$$T_{BH} = \frac{n+1}{4\pi r_H}. \tag{3.39}$$

As expected, the temperature of the black hole decreases as its size increases, but another key feature of the temperature (3.39) is that it increases as the number of extra dimensions n increases. This has important consequences for the Hawking radiation.

For each value of n, the power spectrum for each type of field has a very similar shape to that for scalar fields from a Schwarzschild ($n = 0$) black hole (see Fig. 3.4 and Harris and Kanti (2003)). As n increases and the temperature increases, the emitted energy in each type of field increases as expected. The comparative emission in scalars (spin-0), fermions (spin-$\frac{1}{2}$) and gauge bosons (photons, spin-1) for different n is shown in Tables 3.1–3.2. In Table 3.1, the total energy emission for each species for each value of n is divided by the emission for that species in the $n = 0$ case, to see how the emission changes as n increases. It can be seen that, as the temperature increases, the total energy emission for each species increases rapidly, particularly

Table 3.1 Ratios of total energy emission on the brane from a non-rotating black hole, compared with the emission in the $n = 0$ case for each species

	$n = 0$	$n = 1$	$n = 2$	$n = 3$	$n = 4$	$n = 5$	$n = 6$	$n = 7$
Scalars	1.0	8.94	36.0	99.8	222	429	749	1220
Fermions	1.0	14.2	59.5	162	352	664	1140	1830
Photons	1.0	27.1	144	441	1020	2000	3530	5740

Data taken from Harris and Kanti (2003)

Table 3.2 Ratios of total energy emission on the brane from a non-rotating black hole, compared with the scalar field emission for each value of n

	$n = 0$	$n = 1$	$n = 2$	$n = 3$	$n = 4$	$n = 5$	$n = 6$	$n = 7$
Scalars	1.0	1.0	1.0	1.0	1.0	1.0	1.0	1.0
Fermions	0.55	0.87	0.91	0.89	0.87	0.85	0.84	0.82
Photons	0.23	0.69	0.91	1.00	1.04	1.06	1.06	1.07

Data taken from Harris and Kanti (2003)

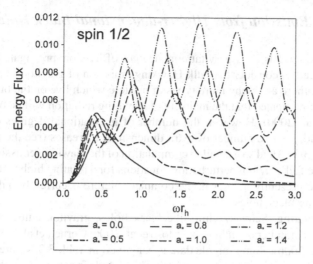

Fig. 3.6 Energy flux (3.32) for spin-$\frac{1}{2}$ particles as a function of ω, for emission on the brane of a higher-dimensional black hole. The number of extra dimensions is fixed to be $n = 1$, and the angular momentum parameter $a_* = a/r_H$ varies. Data taken from Casals et al. (2007)

for spin-1 particles. In Table 3.2, the total energy emission for each species for each value of n is divided by the scalar emission for that value of n, to see how the ratios of the different types of particle emitted change as n changes. The proportion of photon emission increases rapidly as n increases, and the proportion of fermion emission also increases, but not as rapidly. For larger values of n, the photon energy emission is roughly the same as the scalar energy emission and the fermion energy emission is a little smaller. These data are for a single species, so in practice the fermion emission will be rather larger due to the number of light fermions.

For a rotating neutral brane black hole, the metric is (2.2) and the temperature given by (3.2). The physics is now more complicated because the angular momentum of the black hole can vary as well as n. The effect of varying a can be seen in Fig. 3.6, which shows the power spectrum for fermion emission from a black hole with $n = 1$, and different values of a. When the black hole is non-rotating $a = 0$, the spectrum has the Planck shape seen for scalar field emission in Fig. 3.4. As a increases, we see peaks in the spectrum (compare Fig. 3.5) as distinct modes become dominant at different values of the mode energy ω. These peaks become more prominent as a increases. It can also be seen from Fig. 3.6 that the total energy emitted increases rapidly as a increases.

As the number of extra dimensions n increases, the temperature (3.2) increases, with increased emission at every energy ω as n increases. For rotating brane black holes, it is found that typically the scalar and fermion fluxes (for one species) are roughly the same order of magnitude, while the photon flux is typically 10 times that for the scalars and fermions.

3.5.3 Bulk Emission from Higher-dimensional Black Holes

In brane world models, only gravitational degrees of freedom propagate in the bulk, so we only need to consider the bulk Hawking emission of gravitons, and possibly scalars. Since there are many more types of particle which live on the brane than in the bulk, it is anticipated that the majority of Hawking radiation will be on the brane (Emparan et al. 2000). However, the number of gravitational degrees of freedom increases rapidly as the number of extra dimensions increases (see the degeneracy factors (3.36–3.37)) and so a detailed computation of the Hawking emission is necessary. Since the full graviton perturbation equations for a rotating higher-dimensional black hole are presently unknown, the complete picture is currently only available for non-rotating black holes.

For non-rotating, Schwarzschild-like black holes, graviton emission has been considered by a number of authors (Cardoso et al. 2006; Cornell et al. 2006; Creek et al. 2006; Park 2006). Some sample data is contained in Table 3.3, where the energy emitted in bulk gravitons from a non-rotating black hole is compared with the energy emitted on the brane in the form of scalar fields. It can be seen that the energy emitted in bulk gravitons is negligible for small numbers of extra dimensions but rapidly increases as the number of extra dimensions increases, due to the additional gravitational degrees of freedom.

To see whether black holes do indeed "radiate mainly on the brane" (Emparan et al. 2000), we need to take into account the number of particle species of different spin. Table 3.4 shows the percentage of energy emitted by different types of particles, assuming that the particle species correspond to the standard model of particle physics with three families (Cardoso et al. 2006). Even with seven extra dimensions, it can

Table 3.3 Comparison of the energy emitted in bulk gravitons and the energy emitted in scalar fields on the brane, for a non-rotating black hole

	$n=0$	$n=1$	$n=2$	$n=3$	$n=4$	$n=5$	$n=6$	$n=7$
Scalars	1	1	1	1	1	1	1	1
Gravitons	0.02	0.2	0.6	0.91	1.9	2.5	5.1	7.6

Data taken from Cardoso et al. (2006)

Table 3.4 Percentages of energy emission from a non-rotating black hole into particles of spin-0, $\frac{1}{2}$, 1 and 2, assuming the standard model of particle physics with three families and one Higgs field on the brane and only graviton emission in the bulk

	$n=0$	$n=1$	$n=2$	$n=3$	$n=4$	$n=5$	$n=6$	$n=7$
Scalars	6.8	4.0	3.7	3.6	3.6	3.5	3.3	2.9
Fermions	83.8	78.7	75.0	72.3	69.9	68.1	61.6	53.4
Gauge bosons	9.3	16.7	20.0	21.7	22.3	22.2	20.7	18.6
Gravitons	0.1	0.6	1.3	2.4	4.2	7.7	14.4	25.1

Data taken from (Cardoso et al. 2006)

be seen that most of the emission is indeed on the brane. This is due to the large number of fermion degrees of freedom in the standard model.

The question of how this scenario changes when the black hole is rotating cannot be fully answered at present. However, the study of scalar field emission in the bulk (Casals et al. 2008) and tensor-type graviton emission (Doukas et al. 2009; Kanti et al. 2009) has provided some partial answers. Considering just a single degree of freedom, a scalar field, it is found in Casals et al. (2008) that increasing the angular momentum of the black hole decreases the proportion of radiation emitted in the bulk compared with on the brane (due to the rotation of the black hole in the brane). Turning to graviton emission, numerical calculations (Doukas et al. 2009; Kanti et al. 2009) reveal that the emission in tensor-type gravitons increases dramatically as n increases, as might be expected due to the rapid increase in the number of degrees of freedom (3.37). For small n, the emission in gravitons is negligible compared with the emission on the brane, but it will dominate for large n. Tensor-type gravitational degrees of freedom dominate for large n (see the degeneracy factors (3.36)) so one might hope that the unknown contributions from scalar- and vector-type graviton modes do not make a huge difference.

3.6 More General Hawking Emission

The previous section considered the emission of massless, uncharged particles from a neutral black hole. In this short review we do not have room for detailed discussion of generalizations to different black hole geometries or brane-world scenarios. Instead we briefly mention the effects of mass and charge.

The emission of massive particles, as might be expected, is suppressed compared to the emission of massless particles because of the additional energy required to emit a more massive particle (Kanti 2012). There is also a cut-off, in that particles having an energy lower than the rest-mass cannot be emitted.

Charge effects are more complex. In the four-dimensional case, the Hawking radiation of charged fermions from a Reissner-Nordström black hole with metric (2.6) was studied in detail by Page (1977). The emission, on the brane, of both neutral and charged particles from a rotating charged brane black hole has been studied more recently by Sampaio (2009, 2010). The addition of charge to the black hole increases its temperature, which enhances the emission of neutral particles. For charged particles, the differential charge flux per unit time and unit energy is (Sampaio 2009, 2010) (note that this is for brane emission so there is no sum over the index j)

$$\frac{d^2 Q}{dt\, d\omega} = \frac{1}{2\pi} \sum_{\ell=s}^{\infty} \sum_{m=-\ell}^{\ell} \frac{q}{e^{\hat{\omega}/T_{BH}} \pm 1} \mathcal{N}_\Lambda \mathcal{T}_\Lambda \qquad (3.40)$$

where $\mathcal{N}_\Lambda$ is the brane degeneracy factor (3.34),

$$\hat{\omega} = \omega - m\Omega_H - q\Phi_H \qquad (3.41)$$

and Φ_H is the electrostatic potential (2.29). The energy in the Planck factor in the fluxes of particles, energy and angular momentum (3.31–3.33) also changes from $\tilde{\omega}$ (3.23) to $\hat{\omega}$ (3.41) due to the charge of the emitted particles. The thermal factor in (3.40) favours the emission of particles having charge of the same sign as that of the black hole, so that the black hole seeks to lose its charge through Hawking radiation, in much the same way as it loses its angular momentum during the "spin-down" phase of evolution. There is also a "charge super-radiance" effect for bosonic fields Sampaio 2009, 2010, where the greybody factor $\mathcal{T}_\Lambda$ becomes negative for modes where $\hat{\omega} < 0$, analogous to the super-radiance effect for rotating black holes, as shown in Fig. 3.3.

3.7 Conclusions

In this chapter we have examined the Hawking radiation of quantum particles from both four-dimensional and higher-dimensional black holes, including the brane black holes described in Sect. 2.4. We have presented all the equations currently available for the computation of Hawking radiation from both rotating and non-rotating black holes. The complete set of equations governing graviton emission from rotating higher-dimensional black holes are presently elusive.

The evolution of an evaporating microscopic black hole is of particular interest. In four space-time dimensions, a rotating black hole sheds its angular momentum very quickly (Page 1976) and the "spin-down" phase (see Sect. 3.3) is very short compared to the "Schwarzschild" phase where the rotation of the black hole can safely be ignored. In higher dimensions, the black hole sheds its angular momentum quickly, but the "spin-down" phase is more important, with the black hole still having a non-negligible rotation parameter a when it has lost half its mass. It has been estimated that more than 70–80 % of a brane black hole's mass can be radiated away during the "spin-down" phase (Ida et al. 2006).

The study of Hawking radiation proceeds in the semi-classical approximation to quantum gravity, where the black hole geometry is regarded as classical and the quantum field a perturbation of this geometry. This approximation is only valid when the energy of a particle emitted in Hawking radiation is small compared to the mass of the black hole. Towards the end of the lifetime of the black hole, this approximation breaks down because the mass of the black hole is very small. In the absence of a full theory of quantum gravity, the evolution of the black hole during its final "Planck" phase remains mysterious.

Hawking's result predicts that a black hole will evaporate away completely, leaving just the (almost precisely thermal) radiation. This raises a fundamental difficulty, known as the *information loss paradox*. We saw in Chap. 2 that a classical black hole destroys virtually all the information about objects which fall down the event horizon. This is not a problem in classical physics. However, in quantum mechanics one of the key precepts is unitarity, namely that information is conserved and not destroyed. If we view the formation and subsequent evaporation of a black hole as a quantum-mechanical process, then nearly all of the information in the initial quantum state (corresponding to the matter which collapses to form the black hole) is lost as the final state is thermal radiation. This paradox illustrates the fundamental incompatibility of classical general relativity and quantum mechanics, which is one of the reasons why a complete theory of quantum gravity is proving so hard to find. See Hossenfelder and Smolin (2010) for a fuller discussion of the information loss paradox and a survey of attempts at a resolution.

References

Birrell, N.D., Davies, P.C.W.: Quantum Fields in Curved Space, Cambridge University Press (1984)
Cardoso, V., Cavaglia M., Gualtieri L.: Phys. Rev. Lett. **96**, 071301 (2006) [Erratum-ibid. **96**, 219902 (2006)].
Cardoso, V., Cavaglia, M., Gualtieri, L.: JHEP **0602**, 021 (2006)
Casals, M., Dolan, S.R., Kanti, P., Winstanley, E.: JHEP **0703**, 019 (2007)
Casals, M., Dolan, S.R., Kanti, P., Winstanley, E.: JHEP **0806**, 071 (2008)
Casanova, A., Spallucci, E.: Class. Quant. Grav. **23**, R45–R62 (2006)
Chandrasekhar, S.: The Mathematical Theory of Black Holes, Oxford University Press (1998)
Cornell, A.S., Naylor, W., Sasaki, M.: JHEP **0602**, 012 (2006)
Creek, S., Efthimiou, O., Kanti, P., Tamvakis, K.: Phys. Lett. B **635**, 39–49 (2006)
Dai, D.-C., Starkman, G., Stojkovic, D., Issever, C., Rizvi, E., Tseng, J.: Phys. Rev. D **77**, 076007 (2008)
Doukas, J., Cho, H.T., Cornell, A.S., Naylor, W.: Phys. Rev. D **80**, 045021 (2009)
Emparan, R., Horowitz, G.T., Myers, R.C.: Phys. Rev. Lett. **85**, 499–502 (2000)
Frost, J.A., Gaunt, J.R., Sampaio, M.O.P., Casals, M., Dolan, S.R., Parker, M.A., Webber, B.R.: JHEP **0910**, 014 (2009)
Fulling, S. A.: Aspects of Quantum Field Theory in Curved Space-Time, Cambridge University Press (1989)
Giddings, S.B., Thomas, S.D.: Phys. Rev. D **65**, 056010 (2002)
Harris, C.M., Kanti, P.: JHEP **0310**, 014 (2003)
Hawking, S.W.: Nature **248**, 30–31 (1974)
Hawking, S.W.: Commun. Math. Phys. **43**, 199–220 (1975)
Hossenfelder, S., Smolin, L.: Phys. Rev. D **81**, 064009 (2010)
Ida D., Oda K.-y., Park S. C.: Phys. Rev. D **73**, 124022 (2006)
Ishibashi, A., Kodama, H.: Prog. Theor. Phys. **110**, 901–919 (2003)
Jacobson, T.: gr-qc/0308048 (2003)
Kanti, P.: Int. J. Mod. Phys. A **19**, 4899–4951 (2004)
Kanti, P.: Lect. Notes Phys. **769**, 387–423 (2009)
Kanti, P., Kodama, H., Konoplya, R.A., Pappas, N., Zhidenko, A.: Phys. Rev. D **80**, 084016 (2009)
Kanti, P.: Rom. J. Phys. **57**, 879–893 (2012)
Kodama, H., Ishibashi, A.: Prog. Theor. Phys. **110**, 701–722 (2003)
Kodama, H.: Prog. Theor. Phys. Suppl. **172**, 11–20 (2008)
Kodama, H.: Lect. Notes Phys. **769**, 427–470 (2009)
Landsberg, G.L.: Eur. Phys. J. C **33**, S927–S931 (2004)
Mukhanov, V., Winitzki, S.: Introduction to Quantum Effects in Gravity, Cambridge University Press (2007)
Page, D.N.: Phys. Rev. D **13**, 198–206 (1976)

Page, D.N.: Phys. Rev. D **14**, 3260–3273 (1976)
Page, D.N.: Phys. Rev. D **16**, 2402–2411 (1977)
Page, D.N.: New J. Phys. **7**, 203 (2005)
Park, D.K.: Phys. Lett. B **638**, 246 (2006)
Parker, L., Toms, D.: Quantum Field Theory in Curved Space-Time: Quantized Fields and Gravity, Cambridge University Press (2009)
Sampaio, M.O.P.: JHEP **0910**, 008 (2009)
Sampaio, M.O.P.: JHEP **1002**, 042 (2010)
Teukolsky, S.A.: Phys. Rev. Lett. **29**, 1114–1118 (1972)
Teukolsky, S.A.: Astrophys. J. **185**, 635–647 (1973)
Wald, R.M.: Quantum Field Theory in Curved Space-Time and Black Hole Thermodynamics, University of Chicago Press (1994)
Wald, R.M.: Living Rev. Rel. **4**, 6 (2001)
Winstanley, E., arXiv:0708.2656 [hep-th] (2007)

Chapter 4
Primordial Black Holes

Abstract Primordial black holes are the most plausible realization of quantum black holes. Although there is no definite evidence for their existence, they could provide a unique probe of the early Universe, high-energy physics, extra dimensions and even quantum gravity. In particular, the many limits on the fraction of the Universe going into evaporating ones in the mass range 10^9–10^{17} g provide important constraints on models of the early Universe. The strongest limits in this range are associated with their effects on big bang nucleosynthesis and the extragalactic photon background. There would also be a strong constraint at lower masses if evaporating black holes leave stable relics and this would have important implications for models of quantum gravity.

4.1 Introduction

Black holes with a wide range of masses could have formed in the early Universe as a result of the great compression associated with the big bang (Zeldovich and Novikov 1967; Hawking 1971; Carr and Hawking 1974). A comparison of the cosmological density at a time t after the big bang with the density associated with a black hole of mass M suggests that such "primordial" black holes (PBHs) would have a mass of order

$$M \sim \frac{c^3 t}{G} \sim 10^{15} \left(\frac{t}{10^{-23} \,\text{s}} \right) \text{g}. \tag{4.1}$$

This roughly corresponds to the particle horizon mass in a non-inflationary model. PBHs could thus span an enormous mass range: those formed at the Planck time (10^{-43} s) would have the Planck mass (10^{-5} g), whereas those formed at 1 s would be as large as $10^5 \, M_\odot$, comparable to the mass of the holes thought to reside in galactic nuclei. By contrast, black holes forming at the present epoch could never be smaller than about $1 \, M_\odot$.

X. Calmet et al., *Quantum Black Holes*, SpringerBriefs in Physics,
DOI: 10.1007/978-3-642-38939-9_4, © The Author(s) 2014

The realization that PBHs might be small prompted Hawking to study their quantum properties. This led to his famous discovery (Hawking 1974, 1975) that black holes radiate thermally with a temperature

$$T_{BH} = \frac{\hbar c^3}{8\pi G M k_B} \sim 10^{-7} \left(\frac{M}{M_\odot}\right)^{-1} \text{K},\qquad(4.2)$$

so they evaporate completely on a timescale

$$\tau(M) \sim \frac{G^2 M^3}{\hbar c^4} \sim 10^{64} \left(\frac{M}{M_\odot}\right)^3 \text{yr}.\qquad(4.3)$$

Only PBHs smaller than $M_* \approx 10^{15}$ g would have evaporated by the present epoch, so Eq. (4.1) implies that this effect could be important only for ones which formed before 10^{-23} s. Since PBHs with a mass of around 10^{15} g would be producing photons with energy of order 100 MeV at the present epoch, the observational limit on the γ-ray background intensity at 100 MeV immediately implied that their density could not exceed about 10^{-8} times the critical density (Page and Hawking 1976). This suggested that there was little chance of detecting their final explosive phase at the present epoch, at least in the Standard Model of particle physics. It also meant that PBHs with an extended mass function could provide the dark matter only if the fraction of their mass around 10^{15} g were tiny. Nevertheless, it was soon realized that the γ-ray background limit does not preclude PBHs having important cosmological effects (Carr 1976).

The plan of this chapter is as follows. Section 4.2 briefly describes the formation mechanisms for PBHs and their possible cosmological consequences. Section 4.3 discusses in more detail their production from inhomogeneities and inflation. Section 4.4 reviews black hole evaporation and the effects of quark-gluon emission. Section 4.5 discusses the constraints on the fraction of the early Universe going into PBHs from cosmological nucleosynthesis effects, while Sect. 4.6 discusses the ones associated with the photon background. Section 4.7 compares these constraints with other ones in the mass range 10^9–10^{17} g and considers ways in which one might obtain more positive evidence for PBHs. The discussion is based in part on the recent paper by Carr et al. (2010), henceforth referred to as CKSY. Our focus will be primarily on evaporating PBHs since these are the only ones which are regarded as "quantum".

4.2 Formation, Consequences and Abundance of PBHs

4.2.1 How PBHs Form

The high density of the early Universe is a necessary but not sufficient condition for PBH formation. Most PBH formation scenarios depend on the development of inhomogeneities of some kind. Overdense regions could then stop expanding and recollapse (Carr 1975). One possibility is that these inhomogeneities were primordial, in the sense that they were fed into the initial conditions of the Universe. But they may also have arisen spontaneously in an initially smooth Universe—for example, through quantum effects during an inflationary epoch. In either case, the fluctuations would need to be large in order to ensure collapse against the pressure. We discuss these scenarios in more detail in Sect. 4.3. Another possibility is that some sort of phase transition may have enhanced PBH formation or triggered it even if there were no prior inhomogeneities. We now discuss these more exotic scenarios briefly.

Soft equation of state. Whatever the source of the inhomogeneities, PBH formation would be enhanced if some phase transitions led to a sudden reduction in the pressure—for example, at the QCD era (Jedamzik and Niemeyer 1999)—or if the early Universe went through a dustlike phase at early times as a result of either being dominated by non-relativistic particles for a period (Khlopov and Polnarev 1980) or undergoing slow reheating after inflation (Khlopov et al. 1985). In such cases, the effect of pressure in stopping collapse is unimportant and the probability of PBH formation just depends upon the fraction of regions which are sufficiently spherical to undergo collapse. For a given spectrum of primordial fluctuations, this means that there may just be a narrow mass range—associated with the period of the soft equation of state—in which the PBHs form.

Collapse of cosmic loops. In the cosmic string scenario, one expects some strings to self-intersect and form cosmic loops. A typical loop will be larger than its Schwarzschild radius by the factor $(G\mu)^{-1}$, where μ is the string mass per unit length. Observations imply that $G\mu$ must be less than of order 10^{-6}. However, as discussed by many authors (Polnarev and Zemboricz 1988; Hawking 1989; Garriga and Sakellariadou 1993; Caldwell and Casper 1996; MacGibbon et al. 1998), there is still a small probability that a cosmic loop will get into a configuration in which every dimension lies within its Schwarzschild radius. This probability depends upon both μ and the string correlation scale. Note that the holes form with equal probability at every epoch, so they should have an extended mass spectrum. Black holes might also form through the collapse of string necklaces (Matsuda 2006; Lake et al. 2009).

Bubble collisions. Bubbles of broken symmetry might arise at any spontaneously broken symmetry epoch and various people have suggested that PBHs could form as a result of bubble collisions (Kodama et al. 1981; Crawford and Schramm 1982; Hawking et al. 1982; Moss 1994). However, this happens only if the bubble formation rate per Hubble volume is finely tuned: if it is much larger than the Hubble rate, the entire Universe undergoes the phase transition immediately and there is not time to form black holes; if it is much less than the Hubble rate, the bubbles are very rare and never collide. The holes should have a mass of order the horizon mass at the phase transition, so PBHs forming at the GUT epoch would have a mass of 10^3 g, those forming at the electroweak unification epoch would have a mass of 10^{28} g, and those forming at the QCD phase transition would have mass of around $1M_\odot$. The production of PBHs from bubble collisons at the end of 1st order inflation has also been studied (Khlopov et al. 2000).

Collapse of domain walls. The collapse of sufficiently large closed domain walls produced at a 2nd order phase transition in the vacuum state of a scalar field, such as might be associated with inflation, could lead to PBH formation (Rubin et al. 2001; Dokuchaev et al. 2004). These PBHs would have a small mass for a thermal phase transition with the usual equilibrium conditions. However, but they could be much larger if one invoked a non-equlibrium scenario. Indeed, they could then span a wide range of masses, with a fractal structure of smaller PBHs clustered around larger ones (Khlopov et al. 2005).

In most of these scenarios, the PBH mass spectrum is expected to be narrow and centred around the mass given by Eq. (4.1) with t corresponding to the time at which the PBH scale reenters the horizon in the inflationary model or to the time of the relevant cosmological phase transition otherwise. However, PBHs may be smaller than the horizon size at formation in some circumstances. For example, PBH formation is an interesting application of "critical phenomena" and this suggests that their spectrum could extend well below the horizon mass (Yokoyama 1998; Green and Liddle 1999). This would also apply for PBHs formed during a dustlike phase (Polnarev and Khlopov 1985). Note that a PBH could not be much larger than the value given by Eq. (4.1) at formation else it would be a separate closed universe rather than part of our Universe (Harada and Carr 2005).

4.2.2 What PBHs Do

PBHs with $M > 10^{15}$ g. These would still survive today and might be detectable by their gravitational effects. Indeed such PBHs would be obvious dark matter candidates. Since they formed at a time when the Universe was radiation-dominated, they should be classified as non-baryonic and so could avoid the constraints on the baryonic density associated with cosmological nucleosynthesis. They would also be dynamically cold at the present epoch and so would be classified as Cold Dark Matter (CDM). In many respects, they would be like (non-baryonic) WIMPs but they would be much more massive and so could also have the sort of dynamical, lensing and gravitational-wave signatures associated with (baryonic) MACHOs. At one stage there seemed to be evidence for MACHOs with $M \sim 0.5\, M_\odot$ from microlensing observations and PBHs formed at the quark-hadron phase transition seemed one possible explanation for this (Jedamzik 1997). The data no longer support this but there are no constraints excluding PBHs in the sublunar range 10^{20} g $< M < 10^{26}$ g (Blais et al. 2003) or intermediate mass range $10^2\, M_\odot < M < 10^4\, M_\odot$ (Saito and Yokoyama 2009) from having an appreciable density. Large PBHs might also influence the development of large-scale structure (Meszaros 1975; Carr 1977; Freese et al. 1983; Afshordi et al. 2003), seed the supermassive black holes thought to reside in galactic nuclei (Carr and Rees 1984; Bean and Maguiejo 2002; Duchting 2004), generate background gravitational waves (Bond and Carr 1986; Nakamura et al. 1997; Ioka et al. 1999; Inoue and Tanaka 2003) or produce X-rays through accretion and thereby affect the thermal history of the Universe (Ricotti et al. 2008).

PBHs with $M \sim 10^{15}$ g. As already noted, these would be evaporating today and, since they are dynamically cold, one would expect some of them to have clustered within the Galactic halo. Besides contributing to the cosmological γ-ray background, such PBHs could contribute to the Galactic γ-ray background (Wright 1996; Lehoucq et al. 2009) and the antiprotons or positrons in cosmic rays (Kiraly et al. 1981). They might also generate gamma-ray bursts (Cline and Hong 1992), radio bursts (Rees 1977) or the annihilation-line radiation coming from centre of the Galaxy (Okele and Rees 1980). The energy distribution of the particles emitted could also give significant information about the high-energy physics involved in the final explosive phase of black hole evaporation (Halzen et al. 1991).

PBHs with $M < 10^{15}$ g. These would have completely evaporated by now but many processes in the early Universe could have been modified by them. For example, PBH evaporations occurring in the first second of the big bang could generate the entropy of the Universe (Zeldovich and Starobinski 1976), change the details of baryogenesis (Dolgov et al. 2000; Bugaev 2003) or cosmological nucleosynthesis and provide a source of neutrinos (Bugaev and Konishchev 2002) or gravitinos (Khlopov et al. 2006) or other hypothetical particles (Lemoine 2003). If the evaporations left stable Planck-mass relics, these might also contribute to the dark matter (MacGibbon 1987; Barrow et al. 1992; Carr et al. 1994; Green and Liddle 1997; Chen and Adler 2003; Barrau et al. 2003; Alexander and Meszaros 2007). PBHs evaporating at later times could also have important astrophysical effects, such as helping to reionize the Universe (He and Fang 2002).

Even if PBHs had none of these effects, it is still important to study them because each one is associated with an interesting upper limit on the fraction of the mass of the Universe which can have gone into PBHs on some mass-scale M. This fraction is epoch-dependent but its value at the formation epoch of the PBHs, denoted by $\beta(M)$, is of great cosmological interest. The limits associated with big bang nucleosynthesis (BBN) and the extragalactic gamma-ray background (EGB) turn out to be the dominant ones over the mass range 10^9–10^{17} g and both of these have been reassessed recently in the light of observational and theoretical developments (Carr et al. 2010). On the observational front, there are new data on the light element abundances and the γ-ray background density. On the theoretical front, QCD theory suggests that hadrons from evaporating PBHs are produced not from direct emission but from the fragmentation of quark and gluon jets (MacGibbon 1991).

4.2.3 Mass and Density Fraction of PBHs

In the following discussion, we assume that the standard ΛCDM model applies, with the age of the Universe being $t_0 = 13.7$ Gyr and the Hubble parameter being $H_0 \equiv 100h$ with $h = 0.72$. We also put $c = \hbar = k_B = 1$ for the rest of this chapter. The Friedmann equation implies that the density ρ and temperature T during the radiation era are given by

$$H^2 = \frac{8\pi G}{3}\rho = \frac{4\pi^3 G}{45}g_* T^4, \tag{4.4}$$

where g_* counts the number of relativistic degrees of freedom. This can be integrated to give

$$t \approx 0.738 \left(\frac{g_*}{10.75}\right)^{-1/2}\left(\frac{T}{1\,\mathrm{MeV}}\right)^{-2}\,\mathrm{s}, \tag{4.5}$$

where g_* and T are normalised to their values at the start of the BBN epoch. Since we are only considering PBHs which form during the radiation era (the ones generated before inflation being diluted to negligible density), the initial PBH mass M is related to the particle horizon mass M_{PH} by

$$M = \gamma\,M_{\mathrm{PH}} = \frac{4\pi}{3}\gamma\,\rho\,H^{-3} \approx 2.03 \times 10^5\,\gamma\left(\frac{t}{1\,\mathrm{s}}\right)M_\odot. \tag{4.6}$$

Here γ is a numerical factor which depends on the details of gravitational collapse and whose likely value is discussed later.

Throughout this chapter we assume that the PBHs all have the same mass M. This simplifies the analysis considerably and suffices providing we only require limits on the PBH abundance at particular values of M. The current density parameter associated with PBHs which form at a redshift z or time t is related to β by

$$\Omega_{\mathrm{PBH}} = \beta\Omega_R(1+z) \sim 10^6\beta\left(\frac{t}{s}\right)^{-1/2} \sim 10^{18}\beta\left(\frac{M}{10^{15}g}\right)^{-1/2}, \tag{4.7}$$

where $\Omega_R \sim 10^{-4}$ is the density parameter of the microwave background and we have used Eq. (4.1). The $(1+z)$ factor arises because the radiation density scales as $(1+z)^4$, whereas the PBH density scales as $(1+z)^3$. A more precise form of this equation is

$$\Omega_{\mathrm{PBH}} \approx \left(\frac{\beta(M)}{1.15\times 10^{-8}}\right)\left(\frac{h}{0.72}\right)^{-2}\gamma^{1/2}\left(\frac{g_{*i}}{106.75}\right)^{-1/4}\left(\frac{M}{M_\odot}\right)^{-1/2}, \tag{4.8}$$

where g_{*i} is the value of g_* at the epoch of PBH formation. This is normalised to its value at around 10^{-5} s since g_* does not increase much before that in the Standard Model and that is the period in which most PBHs are likely to form. Note that the relationship between β and Ω_{PBH} must be modified if the Universe ever deviates from the standard radiation-dominated behaviour—for example, if there is a dust-like stage for some extended early period or a second inflationary phase or if there are extra dimensions (Sendouda 2005) or if the gravitational constant varies (Barrow 1992; Barrow and Carr 1996; Harada et al. 2002).

Any limit on Ω_{PBH} places a constraint on $\beta(M)$. For non-evaporating PBHs with $M > 10^{15}$g, one constraint comes from requiring that Ω_{PBH} be less than the CDM density, $\Omega_{\mathrm{CDM}} = 0.11h^{-2}$, which implies

$$\beta(M) < 2.04 \times 10^{-18} \, \gamma^{-1/2} \left(\frac{\Omega_{CDM}}{0.25}\right) \left(\frac{h}{0.72}\right)^2 \left(\frac{g_{*i}}{106.75}\right)^{1/4} \left(\frac{M}{10^{15} \, g}\right)^{1/2}.$$
$$\text{(4.9)}$$

Since β always appears in combination with $\gamma^{1/2} \, g_{*i}^{-1/4}$, it is convenient to define a new parameter

$$\beta'(M) \equiv \gamma^{1/2} \left(\frac{g_{*i}}{106.75}\right)^{-1/4} \beta(M), \qquad \text{(4.10)}$$

where g_{*i} can be specified very precisely but γ is rather uncertain. Most of the constraints discussed in this chapter will be expressed in terms of β' rather than β.

Much stronger constraints are associated with PBHs smaller than 10^{15} g since they would have evaporated by now. For example, the γ-ray limit implies $\beta(10^{15} \, g) \lesssim 10^{-26}$ and this is the strongest constraint on β over all mass ranges. Other ones are associated with the generation of entropy and modifications to the cosmological production of light elements. There are also constraints below 10^6g based on the (uncertain) assumption that evaporating PBHs leave stable Planck mass relics, an issue which is discussed later.

The constraints on $\beta(M)$ were first brought together by Novikov et al. (1979). An updated version of the constraints was later provided by Carr et al. (1994) and is shown in Fig. 4.1. Subsequently, this diagram has frequently been revised as the relevant effects have been studied in greater detail. One recent version comes from Josan et al. (2008) but the most comprehensive version is probably that of CKSY, which is presented later. The important qualitative point is that the value of $\beta(M)$ must be tiny over almost every mass range, even if the PBH density is large today, so any cosmological model which would entail an appreciable fraction of the Universe going into PBHs is immediately excluded.

Although we have assumed a monochromatic mass spectrum, there are some circumstances in which the spectrum would be extended and this means that the

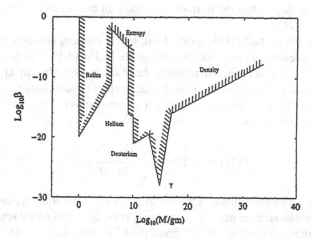

Fig. 4.1 Constraints on $\beta(M)$, from Carr et al. (1994)

constraint on one mass-scale would also imply a constraint on neighbouring scales. For example, the monochromatic assumption fails badly if PBHs form through critical collapse and this modifies the form of the $\beta(M)$ constraint. Another point is that if the PBHs with $M \approx M_*$ have a spread of masses $\Delta M \sim M_*$, one would expect evaporation to lead to a residual spectrum with $n_{PBH} \propto M^3$ for $M < M_*$. This is discussed in more detail by CKSY.

4.3 PBHs as a Probe of Inhomogeneities

One of the most important reasons for studying PBHs is that it enables one to place limits on the spectrum of inhomogeneities in the early Universe. This is because, if the PBHs form directly from density perturbations, the fraction of regions undergoing collapse at any epoch is determined by the root-mean-square amplitude ε of the fluctuations entering the horizon at that epoch and the equation of state $p = \gamma \rho$ ($0 < \gamma < 1$). One usually expects a radiation equation of state ($\gamma = 1/3$) in the early universe but it may have deviated from this in some periods. As we will see, this in turn places constraints on inflationary scenarios, since all of these generate density fluctuations whose spectrum is determined by the form of the inflaton potential.

4.3.1 Simplistic Analysis

Early calculations assumed that the overdense region which evolves to a PBH is spherically symmetric and part of a closed Friedmann model. In order to collapse against the pressure, such a region must be larger than the Jeans length at maximum expansion and this is just $\sqrt{\gamma}$ times the particle horizon size. On the other hand, it cannot be much larger than the horizon size and still be part of our Universe (Carr and Hawking 1974).

This has two important implications. First, PBHs forming at time t after the Big Bang should have of order the horizon mass, given by Eq. (4.1). Second, for a region destined to collapse to a PBH, one requires the fractional overdensity at the horizon epoch δ to exceed γ. Providing the density fluctuations have a Gaussian distribution and are spherically symmetric, one infers that the fraction of regions of mass M which collapse is (Carr 1975)

$$\beta(M) \sim \varepsilon(M) \exp\left[-\frac{\gamma^2}{2\varepsilon(M)^2}\right], \qquad (4.11)$$

where $\varepsilon(M)$ is the rms amplitude of the fluctuations when the horizon mass is M. The PBHs can have an extended mass spectrum only if the fluctuations are scale-invariant (i.e. with ε independent of M). In this case, the PBH mass distribution is given by

$$dn/dM = (\alpha - 2)(M/M_*)^{-\alpha} M_*^{-2} \Omega_{PBH} \rho_{crit}, \qquad (4.12)$$

where Ω_{PBH} is the *total* PBH density parameter and the exponent α is determined by the equation of state:

$$\alpha = \left(\frac{1+3\gamma}{1+\gamma}\right) + 1. \qquad (4.13)$$

$\alpha = 5/2$ if one has a radiation equation of state. This means that the density of PBHs larger than M falls off as $M^{-1/2}$, so most of the PBH density is contained in the smallest ones with mass M_*.

Many scenarios for the cosmological density fluctuations predict that ε is at least approximately scale-invariant but the sensitive dependence of β on ε that means that even tiny deviations from scale-invariance can be important. If $\varepsilon(M)$ decreases with increasing M, then the spectrum falls off exponentially and most of the PBH density is contained in the smallest ones. If $\varepsilon(M)$ increases with increasing M, the spectrum rises exponentially and PBHs could only form at large scales. However, the CMB anisotropies would then be larger than observed, so this is unlikely.

The constraints on $\beta(M)$ in Fig. 4.1 can be converted into constraints on $\varepsilon(M)$ using Eq. (4.11) and these are shown in Fig. 4.2. Also shown is the (non-PBH) constraint associated with the spectral distortions in the CMB induced by the dissipation of intermediate scale density perturbations. This shows that one needs the fluctuation amplitude to decrease with increasing scale in order to produce PBHs and lines corresponding to various slopes in the $\varepsilon(M)$ relationship are also shown in Fig. 4.2.

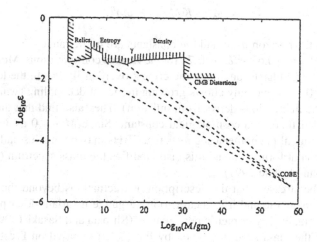

Fig. 4.2 Constraints on $\varepsilon(M)$, from Carr et al. (1994)

4.3.2 Refinements of Simplistic Analysis

The simple criterion for PBH formation given above needs to be tested with detailed numerical calculations. The first hydrodynamical studies of PBH formation (Nadezhin et al. 1978) roughly confirmed the $\delta > \gamma$ criterion for PBH formation, although the PBHs were found to be somewhat smaller than the horizon. Later several groups carried out more detailed hydrodynamical calculations (Niemeyer and Jedamzik 1999; Shibata and Sasaki 1999) and these refined the $\delta > \gamma$ criterion, suggesting that one needs $\delta > 0.7$ for $\gamma = 1/3$ rather than $\delta > 0.3$ and affecting the estimate for $\beta(M)$ given by Eq. (4.11).

A particularly interesting development has been the application of "critical phenomena" to PBH formation. Studies of the collapse of various types of spherically symmetric matter fields have shown that there is always a critical solution which separates those configurations which form a black hole from those which disperse to an asymptotically flat state (Choptuik 1993). The configurations are described by some index p and, as the critical index p_c is approached, the black hole mass is found to scale as $(p - p_c)^\eta$ for some exponent η. This effect was first discovered for scalar fields but subsequently demonstrated for more general fluids with $p = \gamma\rho$.

In all these studies the spacetime was assumed to be asymptotically flat. However, Niemeyer and Jedamzik (1998) applied the same idea to study black hole formation in asymptotically Friedmann models and found similar results. For a variety of initial density perturbation profiles, they found that the relationship between the PBH mass and the horizon-scale density perturbation has the form

$$M = K M_H (\delta - \delta_c)^\eta, \tag{4.14}$$

where M_H is the horizon mass and the constants are in the range $0.34 < \eta < 0.37$, $2.4 < K < 11.9$ and $0.67 < \delta_c < 0.71$ for the various configurations. More recently, Musco et al. (2005) have found that the critical overdensity lies in the lower range $0.43 < \delta_c < 0.47$ if one only allows growing modes at decoupling (which is more plausible if the fluctuations derive from inflation). They also find that the exponent η is modified if there is a cosmological constant. Since $M \to 0$ as $\delta \to \delta_c$, the existence of critical phenomena suggests that PBHs may be much smaller than the particle horizon at formation and this also modifies the mass spectrum (Yokoyama 1998; Green and Liddle 1999).

It should be stressed that the description of fluctuations beyond the horizon is somewhat problematic and it is clearer to use a gauge-invariant description which involves the total energy or metric perturbation (Shibata and Sasaki 1999). Also the derivation of the mass spectrum given by Eq. (4.12) is based on Press-Schechter theory and it is more satisfactory to use peaks theory. Both these points have been considered by Green et al. (2004). They find that the critical value for the density contrast is around 0.3 for $\gamma = 1/3$, which is close to the value originally advocated 30 years ago!

Another refinement of the simplistic analysis which underlies Eq. (4.11) concerns the assumption that the fluctuations have a Gaussian distribution. So long as the fluctuations are small, as certainly applies on a galactic scale, this assumption is reasonable. However, for PBH formation one requires the fluctuations to be large and the coupling of different Fourier modes may then destroy the Gaussianity. An analysis of the fluctuations generated during inflationary (Bullock and Primack 1997; Ivanov 1998; Hidalgo 2007; Byrnes et al. 2012) suggests that $\beta(M)$ can be very different from the form indicated by Eq. (4.11) but it still depends very sensitively on ε.

4.3.3 PBHs and Inflation

Inflation has two important consequences for PBHs (Carr and Lidsey 1993). On the one hand, any PBHs formed before the end of inflation will be diluted to a negligible density. Inflation thus imposes a lower limit on the PBH mass spectrum:

$$M > M_{min} = M_P (T_R / T_P)^{-2}, \qquad (4.15)$$

where T_R is the reheat temperature and $T_P \approx 10^{19}$ GeV is the Planck temperature. The CMB quadrupole measurement implies $T_R \approx 10^{16}$ GeV, so M_{min} certainly exceeds 1 g. On the other hand, inflation will itself generate fluctuations and these may produce PBHs after reheating. If the inflaton potential is $V(\phi)$, then the horizon-scale fluctuations for a mass-scale M are

$$\varepsilon(M) \approx \left(\frac{V^{3/2}}{M_P^3 V'} \right)_H, \qquad (4.16)$$

where a prime denotes $d/d\phi$ and the right-hand side is evaluated for the value of ϕ when the mass-scale M falls within the horizon. In the chaotic inflationary scenario, one makes the "slow-roll" and "friction-dominated" assumptions:

$$\xi \equiv (M_P V'/V)^2 \ll 1, \quad \eta \equiv M_P^2 V''/V \ll 1. \qquad (4.17)$$

The exponent n characterizes the power spectrum of the fluctuations, $|\delta_k|^2 \propto k^n$, and is given by

$$n = 1 - 3\xi + 2\eta \approx 1. \qquad (4.18)$$

This is close to but slightly below 1. Since ε scales as $M^{(1-n)/4}$, this means that the fluctuations are slightly increasing with scale, corresponding to a "red" spectrum. The normalization required to explain galaxy formation ($\varepsilon \approx 10^{-5}$) would then preclude the formation of PBHs on a smaller scale. If PBH formation is to occur, one needs the fluctuations to decrease with increasing mass ($n > 1$), corresponding to a "blue"

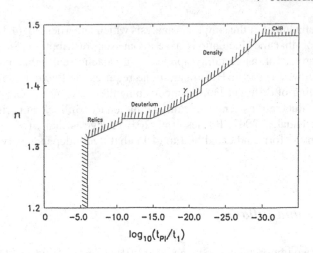

Fig. 4.3 Constraints on spectral index n in terms of reheat time t_1

spectrum, and from Eq. (4.18) this is only possible if the scalar field is accelerating sufficiently fast that $V''V/V'^2 > 3/2$. This condition is certainly satisfied in some scenarios (Gilbert 1995) and, if it is, Eq. (4.11) implies that the PBH density will be dominated by the ones forming immediately after reheating. This is because the volume dilution of the PBHs forming shortly before the end of inflation will dominate the enhancement associated with Eq. (4.11). However, it should be stressed that the validity of Eq. (4.16) at the very end of inflation is questionable since the usual assumptions may fail then (Lyth et al. 2006).

Since each value of n corresponds to a straight line in Fig. 4.3, any particular value for the reheat time t_1 corresponds to an upper limit on n. This limit is indicated in Fig. 4.3, which is taken from Carr et al. (1994) but incorporates a correction from Green and Liddle (1997). At the time this was the strongest limit on n available. Nowadays there are stronger constraints from the CMB anisotropies and these imply $n < 1$ on large scales. Hence PBHs could form only if n is scale-dependent and exceeds 1 below some scale. In this case, Fig. 4.3 is inapplicable but can be adapted if n is constant below this scale.

It should be stressed that not all inflationary scenarios predict that the spectral index should be constant. In some scenarios, the fluctuations have a "running index", so that the amplitude will not increase on smaller scales according to a simple power law. In others, the potential is flattened over some range and Eq. (4.16) then implies that there is a spike in the spectrum. Indeed, many people have invoked some form of "designer" inflation, in which the power spectrum of the fluctuations—and hence PBH production—peaks on some scale (Hodges and Blumenthal 1990; Yokoyama 1997, 1998, 1999). For example, one can fine-tune the position of the spike so that it corresponds to the mass-scale associated with the microlensing events observed towards the Large Magellanic Cloud (Ivanov et al. 1994). Besides the chaotic scenario, there are numerous variants of inflation (supernatural, supersymmetric, hybrid, multiple, oscillating, ghost, running mass, saddle, hilltop etc.) and PBH formation has been studied in all of these models. Full references can be found in CKSY. So even if PBHs never actually formed, studying them places important constraints on the many types of inflationary scenarios.

Note that in the standard scenario inflation ends by the decay of the inflaton into radiation. However, in the preheating scenario it ends more rapidly because of resonant coupling between the inflaton and another scalar field. This generates extra fluctuations, which are not of the form indicated by Eq. (4.16) but might also produce PBHs (Green and Malik 2001; Suyama et al. 2005). Such fluctuations peak on a scale associated with reheating. This is usually very small but several scenarios involve a secondary inflationary phase which boosts this scale into the macroscopic domain (Saito and Yokoyama 2009). There are also other scenarios for generating perturbations at the end of inflation (Bernardeau et al. 2004), in which the probability of PBH formation is again be unrelated to Eq. (4.16).

4.4 Evaporation of Primordial Black Holes

4.4.1 Lifetime

As discussed in Chap. 3, a black hole with mass $M \equiv M_{10} \times 10^{10}$ g emits thermal radiation with temperature

$$T_{\mathrm{BH}} = \frac{1}{8\pi \, G \, M} \approx 1.06 \, M_{10}^{-1} \, \mathrm{TeV}. \qquad (4.19)$$

This assumes that the hole has no charge or angular momentum, which is reasonable since charge and angular momentum will be lost through quantum emission on a shorter timescale than the mass. We have seen that the emission is not exactly black-body but depends upon the spin and charge of the emitted particle, the average energy for neutrinos, electrons and photons being $4.22 \, T_{\mathrm{BH}}$, $4.18 \, T_{\mathrm{BH}}$ and $5.71 \, T_{\mathrm{BH}}$, respectively.

The mass loss rate of an evaporating black hole can be expressed as

$$\frac{\mathrm{d}M_{10}}{\mathrm{d}t} = -5.34 \times 10^{-5} \, f(M) \, M_{10}^{-2} \, \mathrm{s}^{-1}. \qquad (4.20)$$

Here $f(M)$ is a measure of the number of emitted particle species, normalised to unity for a black hole with $M \gg 10^{17}$ g, this emitting only particles which are (effectively) massless: photons, three generations of neutrinos and antineutrinos, and gravitons. The contribution of each relativistic degree of freedom to $f(M)$ is (MacGibbon 1991)

$$f_{s=0} = 0.267, \quad f_{s=1} = 0.060, \quad f_{s=3/2} = 0.020, \quad f_{s=2} = 0.007,$$
$$f_{s=1/2} = 0.147 \text{ (neutral)}, \quad f_{s=1/2} = 0.142 \text{ (charge } \pm e\text{)}. \tag{4.21}$$

Holes in the mass range 10^{15} g $< M < 10^{17}$ g emit electrons but not muons, while those in the range 10^{14} g $< M < 10^{15}$ g also emit muons, which subsequently decay into electrons and neutrinos. The latter range is relevant for the PBHs which are completing their evaporation at the present epoch.

Once M falls to around 10^{14} g, a black hole can also begin to emit hadrons. However, hadrons are composite particles made up of quarks held together by gluons. For temperatures exceeding the QCD confinement scale, $\Lambda_{QCD} = 250$–300 MeV, one would expect these fundamental particles to be emitted rather than composite particles. Only pions would be light enough to be emitted below Λ_{QCD}. Above this temperature, the particles radiated can be regarded as asymptotically free, leading to the emission of quarks and gluons. Since there are 12 quark degrees of freedom per flavour and 16 gluon degrees of freedom, one would expect the emission rate (i.e., the value of f) to increase suddenly once the QCD temperature is reached. If one includes just u, d and s quarks and gluons, Eq. (4.21) implies that their contribution to f is $3 \times 12 \times 0.14 + 16 \times 0.06 \approx 6$, compared to the pre-QCD value of about 2. Thus the value of f roughly quadruples, although there will be a further increase in f at somewhat higher temperatures due to the emission of the heavier quarks. After their emission, quarks and gluons fragment into further quarks and gluons until they cluster into the observable hadrons when they have travelled a distance $\Lambda_{QCD}^{-1} \sim 10^{-13}$ cm. This is much larger than the size of the hole, so gravitational effects can be neglected.

If we sum up the contributions from all the particles in the Standard Model up to 1 TeV, corresponding to $M_{10} \sim 1$, this gives $f(M) = 15.35$. Integrating the mass loss rate over time gives a lifetime

$$\tau \approx 407 \left(\frac{f(M)}{15.35} \right)^{-1} M_{10}^3 \text{ s}. \tag{4.22}$$

The mass of a PBH evaporating at time τ after the big bang is then

$$M \approx 1.35 \times 10^9 \left(\frac{f(M)}{15.35} \right)^{1/3} \left(\frac{\tau}{1 \text{ s}} \right)^{1/3} \text{ g}. \tag{4.23}$$

The critical mass for which τ equals the age of the Universe is denoted by M_*. For the currently favoured age of 13.7 Gyr, one finds

$$M_* \approx 1.02 \times 10^{15} \left(\frac{f_*}{15.35} \right)^{1/3} \text{ g} \approx 5.1 \times 10^{14} \text{ g}, \tag{4.24}$$

where the last step assumes $f_* = 1.9$, the value associated with the temperature $T_{\text{BH}}(M_*) = 21$ MeV. At this temperature muons and some pions are emitted, so the value of f_* accounts for this. Although QCD effects are initially small for PBHs with $M = M_*$, only contributing a few percent, it should be noted that they become important once M falls to

$$M_q \approx 0.4 M_* \approx 2 \times 10^{14}\,\mathrm{g}, \qquad (4.25)$$

since the peak energy becomes comparable to Λ_{QCD} then. This means that an appreciable fraction of the time-integrated emission from the PBHs evaporating at the present epoch goes into quark and gluon jet products.

It should be stressed that the above analysis is not exact because the value of $f(M)$ in Eq. (4.23) should really be the weighted average of $f(M)$ over the lifetime of the black hole. The more precise calculation of MacGibbon (1991) gives the slightly smaller value $M_* = 5.00 \times 10^{14}\,\mathrm{g}$. However, the weighted average is well approximated by $f(M)$ unless one is close to a particle mass threshold. For example, since the lifetime of a black hole of mass $0.4\,M_*$ is roughly $0.25 \times (0.4)^3 = 0.016$ that of an M_* black hole, one expects the value of M_* to be overestimated by a few percent. This explains the small difference from MacGibbon's calculation.

4.4.2 Particle Spectra

Particles injected from PBHs have two components: the *primary* component, which is the direct Hawking emission, and the *secondary* component, which comes from the decay of gauge bosons and the hadrons produced by fragmentation of primary quarks and gluons. For example, the photon spectrum can be written as

$$\frac{\mathrm{d}\dot{N}_\gamma}{\mathrm{d}E_\gamma}(E_\gamma, M) = \frac{\mathrm{d}\dot{N}_\gamma^{\mathrm{pri}}}{\mathrm{d}E_\gamma}(E_\gamma, M) + \frac{\mathrm{d}\dot{N}_\gamma^{\mathrm{sec}}}{\mathrm{d}E_\gamma}(E_\gamma, M), \qquad (4.26)$$

with similar expressions for other particles. In order to treat QCD fragmentation, CKSY use the PYTHIA code, a Monte Carlo event generator constructed to fit hadron fragmentation for centre-of-mass energies $\sqrt{s} \lesssim 200\,\mathrm{GeV}$. Similar results are obtained by HERWIG, the code used by MacGibbon and Webber (1990).

The spectrum of secondary photons is peaked around $E_\gamma \simeq m_{\pi^0}/2 \approx 68\,\mathrm{MeV}$, because it is dominated by the 2γ-decay of soft neutral pions which are practically at rest. The peak flux can be expressed as

$$\frac{\mathrm{d}\dot{N}_\gamma^{\mathrm{sec}}}{\mathrm{d}E_\gamma}(E_\gamma = m_{\pi^0}/2) \simeq 2 \sum_{i=q,g} \mathscr{B}_{i \to \pi^0}(\overline{E}, E_{\pi^0}) \frac{\overline{E}}{m_{\pi^0}} \frac{\mathrm{d}\dot{N}_i^{\mathrm{pri}}}{\mathrm{d}E_i}(E_i \simeq \overline{E}), \quad (4.27)$$

where $\mathscr{B}_{q,g \to \pi^0}(E_{\mathrm{jet}}, E_{\pi^0})$ is the fraction of the jet energy E_{jet} going into neutral pions of energy E_{π^0}. This is of order 0.1 and fairly independent of jet energy. If we assume that most of the primary particles have the average energy $\overline{E} \approx 4.4\,T_{\mathrm{BH}}$,

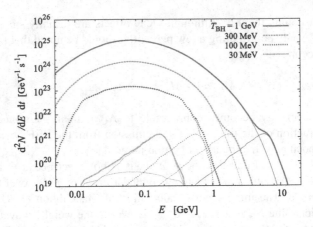

Fig. 4.4 Instantaneous emission rate of photons for four typical black hole temperatures, from CKSY. For each temperature, the curve with the peak to the *right* (*left*) represents the primary (*secondary*) component and the *thick curve* denotes their sum

the last factor becomes $d\dot{N}_i^{\mathrm{pri}}/dE_i \approx 1.6 \times 10^{-3}$. Thus the energy dependence of Eq. (4.27) comes entirely from the factor $\overline{E}$ and is proportional to the Hawking temperature. The emission rates of primary and secondary photons for four typical temperatures are shown in Fig. 4.4.

It should be noted that the time-integrated ratio of the secondary flux to the primary flux drops rapidly once M goes above M_*. This is because a black hole with $M = M_*$ will emit quarks efficiently once its mass gets down to the value M_q given by Eq. (4.25) and this corresponds to an appreciable fraction of its original mass. On the other hand, a PBH with somewhat larger initial mass, $M = (1+\mu) M_*$, will today have a mass

$$m \equiv M(t_0) \approx (3\,\mu)^{1/3} M_* \quad (\mu \ll 1)\,. \tag{4.28}$$

Here we have assumed $f(M) \approx f_*$, which should be a good approximation for $m > M_q$ since the value of f only changes slowly above the QCD threshold. However, m falls below M_q for $\mu < 0.02$ and if we assume that f jumps discontinuously from f_* to αf_* at this mass, then Eq. (4.28) must be reduced by a factor $\alpha^{1/3}$. The fact that this happens only for $\mu < 0.02$ means that the fraction of the black hole mass going into secondaries falls off sharply above M_*. The ratio of the secondary to primary peak energies and the ratio of the time-integrated fluxes are shown in Fig. 4.5.

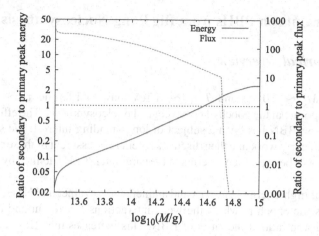

Fig. 4.5 Ratios of secondary to primary peak energies (*solid*) and fluxes (*dashed*), from CKSY

4.4.3 Photosphere Effects

There has been some dispute in the literature about the interactions between emitted particles around an evaporating black hole. The usual assumption that there is no interaction between emitted particles has been refuted by Heckler (1997), who claims that QED interactions could produce an optically thick photosphere once the black hole temperature exceeds $T_{BH} = 45$ GeV. He has proposed that a similar effect may operate at an even lower temperature, $T_{BH} \approx 200$ MeV, due to QCD effects (Heckler 1997). Variants of these models and their astrophysical implications have been studied by various authors (Cline et al. 1999; Kapusta 2001; Daghigh and Kapusta 2006). However, MacGibbon et al. (2008) have identified a number of physical and geometrical effects which invalidate these claims. First, the particles must be causally connected in order to interact and this means that the standard cross-sections are reduced (viz. the particles are created at a finite time and do not go back to the infinite past). Second, because of the Landau–Pomeranchuk–Migdal effect, a scattered particle requires a minimum distance to complete each bremsstrahlung interaction, with the consequence that there is unlikely to be more than one complete bremsstrahlung interaction per particle near the black hole.

MacGibbon et al. conclude that the emitted particles do not interact sufficiently to form a QED photosphere and that the conditions for QCD photosphere formation could only be temporarily satisfied (if at all) when the black hole temperature is of order Λ_{QCD}. Even in this case, the strong damping of the Hawking production of QCD particles around this threshold may suffice to suppress it. In any case, no QCD photosphere persists once the black hole temperature climbs above Λ_{QCD}. They also consider the suggestion that plasma interactions between emitted particles could form a photosphere (Belyanin 1996) but conclude that this too is implausible. In what follows, we therefore assume that no photosphere forms.

4.5 Constraints on PBHs from Big Bang Nucleosynthesis

4.5.1 Historical Overview

PBHs with $M \sim 10^{10}$ g and $T_{BH} \sim 1$ TeV have a lifetime $\tau \sim 10^3$ s and therefore evaporate at the epoch of cosmological nucleosynthesis. The effect of these evaporations on BBN has been a subject of long-standing interest. We start with a brief review of early work and then discuss a recent reassessment of these constraints. All the limits will be expressed in terms of the parameter $\beta'(M)$ defined by Eq. (4.10).

- Injection of high-energy neutrinos and antineutrinos (Vainer and Nasel'skii 1978). This changes the epoch at which the weak interactions freeze out and thereby the neutron-to-proton ratio at the onset of BBN. This increases the ^{4}He production, so demanding that the primordial abundance satisfy $Y_p < 0.33$, the most conservative constraint available at the time, gave a limit

$$\beta'(M) < 3 \times (10^{-18} - 10^{-15}) M_{10}^{1/2} \quad (M = 10^9 - 3 \times 10^{11} \text{ g}). \quad (4.29)$$

- Entropy generation (Miyama and Sato 1978). Since PBHs with $M = 10^9 - 10^{13}$ g evaporated during or after BBN, the baryon-to-entropy ratio at nucleosynthesis would be increased, resulting in overproduction of ^{4}He and underproduction of D. Demanding that the primordial mass fractions of these elements satisfy $Y_p < 0.29$ and $D_p > 1 \times 10^{-5}$ led to a limit

$$\beta'(M) < 10^{-15} M_{10}^{-5/2} \quad (M = 10^9 - 10^{13} \text{ g}). \quad (4.30)$$

- Emission of high-energy nucleons and antinucleons (Zeldovich et al. 1977). The change in the primordial ^{4}He and D abundances due to capture of free neutrons by protons and spallation of ^{4}He gave the upper limits:

$$\beta'(M) < \begin{cases} 6 \times 10^{-18} M_{10}^{-1/2} & (M = 10^9 - 10^{10} \text{ g}), \\ 6 \times 10^{-22} M_{10}^{-1/2} & (M = 10^{10} - 10^{11} \text{ g}), \\ 3 \times 10^{-21} M_{10}^{-1/2} & (M = 10^{11} - 10^{13} \text{ g}). \end{cases} \quad (4.31)$$

- Emission of photons (Lindley 1980). The dissociation of deuterons produced in nucleosynthesis by photons from evaporating PBHs with $M > 10^{10}$ g led to the constraint

$$\beta'(M) < 3 \times 10^{-20} M_{10}^{1/2} \quad (M > 10^{10} \text{ g}). \quad (4.32)$$

This is comparable to the limit from the extra production of deuterons discussed above.

The strongest of the above limits are shown in Fig. 4.1. Observational data on both the light element abundances and the neutron lifetime have changed since these early papers. Much more significant, however, have been developments in our understanding of the fragmentation of quark and gluon jets into hadrons. Most of the hadrons created decay almost instantaneously compared to the timescale of nucleosynthesis, but long-lived ones (such as pions, kaons, and nucleons) remain long enough in the ambient medium to leave an observable signature on BBN. These effects were first discussed by Kohri and Yokoyama (1999) for the relatively low mass PBHs evaporating in the early stages of BBN but the analysis has now been extended to incorporate the effects of heavier PBHs, evaporating after BBN, the hadrons and high energy photons from these PBHs further dissociating synthesised light elements.

4.5.2 Revised Constraints on $\beta'(M)$ Imposed by BBN

High energy particles emitted by PBHs modify the standard BBN scenario in three different ways: (1) high energy mesons and antinucleons induce extra interconversion between background protons and neutrons even after the weak interaction has frozen out in the background Universe; (2) high energy hadrons dissociate light elements synthesised in BBN, thereby reducing ^{4}He and increasing D, T, ^{3}He, ^{6}Li and ^{7}Li; (3) high energy photons generated in the cascade further dissociate ^{4}He to increase the abundance of lighter elements even more.

The PBH constraints depend on three parameters: the initial baryon-to-photon ratio η_i, the PBH initial mass M or (equivalently) its lifetime τ, and the initial PBH number density normalised to the entropy density, $Y_{PBH} \equiv n_{PBH}/s$. From Eq. (4.7) this is related to the initial mass fraction β' by

$$\beta' = 5.4 \times 10^{21} \left(\frac{\tau}{1\,s}\right)^{1/2} Y_{PBH}. \tag{4.33}$$

The parameters β', τ and Y_{PBH} all depend on M but we suppose a monochromatic mass function in what follows. The initial baryon-to-photon ratio is set to the present one, $\eta = (6.225 \pm 0.170) \times 10^{-10}$, after allowing for entropy production from PBH evaporations and photon heating due to e^+e^- annihilations.

Figure 4.6 summarises the results of these calculations. PBHs with lifetime smaller than 10^{-2} s are free from BBN constraints because they evaporate well before weak freeze-out and leave no trace. PBHs with $M = 10^9$–10^{10} g and lifetime $\tau = 10^{-2}$–10^2 s are constrained by the extra interconversion between protons and neutrons due to emitted mesons and antinucleons, which increases the n/p freeze-out ratio as well as the final ^{4}He abundance. For $\tau = 10^2$–10^7 s, corresponding to $M = 10^{10}$–10^{12} g, hadrodissociation processes become important and the debris deuterons and non-thermally produced ^{6}Li put strong constraints on $\beta(M)$. Finally, for $\tau = 10^7$–10^{12} s, corresponding to $M = 10^{12}$–10^{13} g, energetic neutrons decay before inducing hadrodissociation. Instead, photodissociation processes are operative and the most stringent constraint comes from overproduction of ^{3}He or D. However, even these effects become insignificant after 10^{12} s.

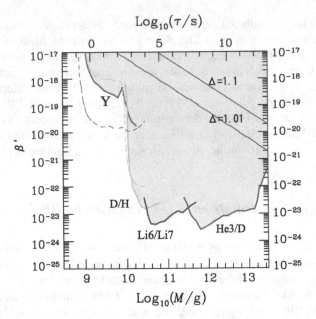

Fig. 4.6 Upper bounds on $\beta'(M)$ from BBN, with broken line giving earlier limit, from CKSY

For comparison, we show the much weaker constraint imposed by the entropy production from evaporating PBHs (Miyama and Sato 1978). The factor Δ in Fig. 4.6 is the ratio of the entropy density after and before PBH evaporations. We also show as a broken line the limits obtained earlier by Kohri and Yokoyama (1999). The helium limit is weaker because the helium abundance is now known to be smaller, while the deuterium limit is stronger because hadrodissociation of helium produces more deuterium.

4.6 Constraints on PBHs from Extragalactic Photon Background

4.6.1 Historical Overview

One of the earliest works that applied the theory of black hole evaporation to astrophysics was carried out by Page and Hawking (1976). They used the diffuse EGB observations to constrain the mean cosmological number density of PBHs which are completing their evaporation at the present epoch to be less than 10^4 pc^{-3}. This corresponds to an upper limit on $\Omega_{\rm PBH}$ of around 10^{-8}. The limit was subsequently refined by MacGibbon and Carr (1991), who considered how it is modified by including quark and gluon emission and inferred $\Omega_{\rm PBH} \leq (7.6 \pm 2.6) \times 10^{-9} h^{-2}$. Later they used EGRET observations to derive a slightly stronger limit $\Omega_{\rm PBH} \leq$

$(5.1 \pm 1.3) \times 10^{-9} h^{-2}$ (Carr and MacGibbon, 1998). Using the modern value of h gives $\Omega_{PBH} \leq (9.8 \pm 2.5) \times 10^{-9}$ and this corresponds to $\beta'(M_*) < 6 \times 10^{-26}$ from Eq. (4.8). They also inferred from the form of the γ-ray spectrum that PBHs could not provide the *dominant* contribution to the background.

4.6.2 Expected EGB from PBHs

The photon emission has a primary and secondary component and these are calculated according to the prescription of Sect. 4.4.2. The relative magnitude of these two components is sensitive to the PBH mass and this affects the associated $\beta'(M)$ limit. In order to determine the present background spectrum of particles generated by PBH evaporations, we must integrate over the lifetime of the black holes, allowing for the fact that particles generated in earlier cosmological epochs will be redshifted in energy by now.

If the PBHs all have the same initial mass M, and if we approximate the number of emitted photons in the energy bin $\Delta E_\gamma \simeq E_\gamma$ by $\dot{N}_\gamma(E_\gamma) \simeq E_\gamma (d\dot{N}_\gamma/dE_\gamma)$, then the emission rate per volume at cosmological time t is

$$\frac{dn_\gamma}{dt}(E_\gamma, t) \simeq n_{PBH}(t) E_\gamma \frac{d\dot{N}_\gamma}{dE_\gamma}(M(t), E_\gamma), \qquad (4.34)$$

where the t-dependence of M just reflects the evaporation. Since the photon energy and density are redshifted by factors $(1+z)^{-1}$ and $(1+z)^{-3}$, respectively, the present number density of photons with energy $E_{\gamma 0}$ is

$$n_{\gamma 0}(E_{\gamma 0}) = n_{PBH0} E_{\gamma 0} \int_{t_{min}}^{min(t_0, \tau)} dt \, (1+z) \frac{d\dot{N}z_\gamma}{dE_\gamma}(M(t), (1+z) E_{\gamma 0}), \qquad (4.35)$$

where t_{min} corresponds to the earliest time at which the photons freely propagate and n_{PBH0} is the current PBH number density for $M > M_*$ or the number density PBHs would have had now had they not evaporated for $M < M_*$. The photon flux is

$$I \equiv \frac{1}{4\pi} n_{\gamma 0}. \qquad (4.36)$$

The calculated present-day fluxes of primary and secondary photons are shown in Fig. 4.7, where the number density n_{PBH0} for each M has the maximum value consistent with the observations.

Note that the highest energy photons are associated with PBHs of mass M_*. Photons from PBHs with $M > M_*$ are at lower energies because they are cooler, while photons from PBHs with $M < M_*$ are at lower energies because (although initially hotter) they are redshifted. The spectral shape depends on the mass M and can be easily understood. Holes with $M > M_*$ have a rather sharp peak which is well approximated by the instantaneous black-body emission of the primary photons, while holes with $M \leq M_*$ have an $E_{\gamma 0}^{-2}$ fall-off for $E_{\gamma 0} \gg T_{BH}/(1 + z(\tau))$ due to the final phases of evaporation (MacGibbon 1991).

4.6.3 Limits on $\beta'(M)$ Imposed by Observed EGB

The relevant observations come from HEAO 1 and other balloon observations in the 3–500 keV range, COMPTEL in the 0.8–30 MeV range, EGRET in the 30–200 MeV range and *Fermi* LAT in the 200 MeV–102 GeV range. All the observations are shown in Fig. 4.7. The origin of the diffuse X-ray and γ-ray backgrounds is thought to be primarily distant astrophysical sources, such as blazars, and in principle one should remove these contributions before calculating the PBH constraints. This is the strategy adopted by Barrau et al. (2003), who thereby obtain a limit $\Omega_{PBH} \lesssim 3.3 \times 10^{-9}$. CKSY do not attempt such a subtraction, so their constraints on $\beta'(M)$ may be overly conservative.

In order to analyse the spectra of photons emitted from PBHs, different treatments are needed for PBHs with initial masses below and above M_*. We saw in Sect. 4.4.2 that PBHs with $M > M_*$ can never emit secondary photons at the present epoch, whereas those with $M \leq M_*$ will do so once M falls below $M_q \approx 2 \times 10^{14}$ g. One can use simple analytical arguments to derive the form of the primary and secondary peak fluxes. The observed X-ray and γ-ray spectra correspond to $I^{obs} \propto E_{\gamma 0}^{-(1+\varepsilon)}$ where ε lies between 0.1 and 0.4. For $M < M_*$, the limit is determined by the secondary flux and one can write the upper bound on β' as

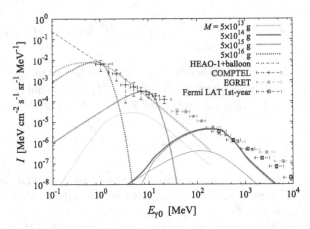

Fig. 4.7 Fluxes corresponding to the upper limit on the PBH abundance for various values of M, from CSKY. All PBHs produce primary photons but $M \lesssim M_*$ ones also produce secondary photons and this gives a stronger constraint on β

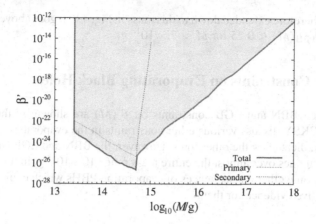

Fig. 4.8 Upper bounds on $\beta'(M)$ from the extragalactic photon background, from CSKY, with no other contributors to the background having been subtracted

$$\beta'(M) \lesssim 3 \times 10^{-27} \left(\frac{M}{M_*}\right)^{-5/2-2\varepsilon} \quad (M < M_*). \qquad (4.37)$$

For $M > M_*$, secondary photons are not emitted and one obtains a limit

$$\beta'(M) \lesssim 4 \times 10^{-26} \left(\frac{M}{M_*}\right)^{7/2+\varepsilon} \quad (M > M_*). \qquad (4.38)$$

These M-dependences explain qualitatively the slopes in Fig. 4.8. The limit bottoms out at 3×10^{-27} and, from Eq. (4.8), the associated limit on the density parameter is $\Omega_{\mathrm{PBH}}(M_*) \lesssim 5 \times 10^{-10}$.

Finally, we determine the mass range over which the γ-ray constraint applies. Since photons emitted at sufficiently early times cannot propagate freely, there is a minimum mass M_{min} below which the above constraint is inapplicable. The dominant interactions between γ-rays and the background Universe in the relevant energy range are pair-production off hydrogen and helium nuclei. For the opacity appropriate for a 75 % hydrogen and 25 % helium mix, the redshift below which there is free propagation is given by (MacGibbon and Carr 1991)

$$1 + z_{\mathrm{max}} \approx 1100 \left(\frac{h}{0.72}\right)^{-2/3} \left(\frac{\Omega_b}{0.05}\right)^{-2/3}, \qquad (4.39)$$

with the nucleon density parameter Ω_b being normalised to the modern value. The condition $\tau(M_{\mathrm{min}}) = t(z_{\mathrm{max}})$ then gives

$$M_{\mathrm{min}} = \left(\frac{t(z_{\mathrm{max}})}{t_0}\right)^{1/3} \left(\frac{f(M_{\mathrm{min}})}{f_*}\right)^{1/3} M_* \approx 3 \times 10^{13} \, \mathrm{g}. \qquad (4.40)$$

The limit is therefore extended down to this mass in Fig. 4.8. It goes above the density constraint $\Omega_{PBH}(M) < 0.25$ for $M > 7 \times 10^{16}$ g.

4.7 Other Constraints on Evaporating Black Holes

The combined BBN and EGB constraints on $\beta'(M)$ are shown by the solid line in Fig. 4.9. CKSY discuss various other constraints in the evaporating mass range, these being indicated by the other lines. However, the BBN and EGB limits are the most stringent ones over almost the entire mass range 10^9–10^{17} g. In this section we will discuss some other consequences of evaporating PBHs which could potentially provide positive evidence for them.

4.7.1 Galactic Gamma-Rays

If PBHs of mass M_* are clustered inside our own Galactic halo, as expected, then there should also be a Galactic γ-ray background and, since this would be anisotropic, it should be separable from the extragalactic background. Some time ago it was claimed that such a background had been detected by EGRET between 30 MeV and 120 GeV and that this could be attributed to PBHs (Wright 1996). A more recent analysis of EGRET data between 70 MeV and 150 GeV, assuming a variety of distributions for the PBHs, gives a limit $\Omega_{PBH}(M_*) \leq 2.6 \times 10^{-9}$ or $\beta'(M_*) < 1.4 \times 10^{-26}$

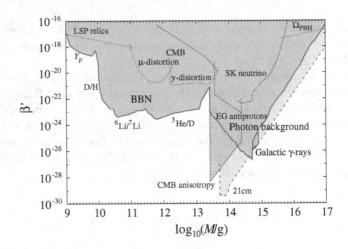

Fig. 4.9 Combined BBN and EGB limits (*solid*), compared to other constraints on evaporating PBHs from LSP relics and CMB distortions (*short-dashed*), extragalactic antiprotons and neutrinos (*dotted*), the Galactic γ-ray background (*long-dashed*), CMB anisotropies (*dash-dotted*), the potential limit from 21 cm observations (*broken*) and the density limit from the smallest unevaporated black holes (*dashed*). From CKSY

(Lehoucq et al. 2009). This is a factor of 5 above the EGB constraint obtained in Sect. 4.6 and corresponds to an explosion rate $\mathscr{R} \leq 0.06\,\mathrm{pc}^{-3}\,\mathrm{yr}^{-1}$. CKSY have analysed this constraint in more detail. Whereas the strongest constraint on $\beta(M)$ from the extragalactic background comes from the time-integrated contribution of the M_* black holes, which peaks at 120 MeV, the Galactic background is dominated by PBHs which are initially slightly larger than this, since the ones with *exactly* the mass M_* no longer exist. Equation (4.28) implies that the emission from PBHs with initial mass $(1 + \mu) M_*$ currently peaks at an energy $E \approx 100\,(3\,\mu)^{-1/3}\,\mathrm{MeV}$, which is in the range 70 MeV–150 GeV for $0.7 > \mu > 0.08$. So these black holes correspond to the $n_{\mathrm{PBH}} \propto M^3$ population.

4.7.2 Galactic Antiprotons

Since the ratio of antiprotons to protons in cosmic rays is less than 10^{-4} over the energy range 100 MeV–10 GeV, whereas PBHs should produce them in equal numbers, PBHs could only contribute appreciably to the antiprotons (Carr 1976). It is usually assumed that the observed antiprotons are secondary particles, produced by spallation of the interstellar medium by primary cosmic rays. However, the spectrum of secondary antiprotons should show a steep cut-off at kinetic energies below 2 GeV, whereas the spectrum of PBH antiprotons should continue down to 0.2 GeV. Also any primary antiproton fraction should tend to 0.5 at low energies. Both these features provide a distinctive signature of any PBH contribution.

The black hole temperature must be much larger than $T_{\mathrm{BH}}(M_*)$ to generate antiprotons, so the local cosmic ray flux from PBHs should be dominated by the ones just entering their explosive phase at the present epoch. Such PBHs should be clustered inside our halo, so any charged particles emitted will have their flux enhanced relative to the extragalactic spectra by a factor ζ which depends upon the halo concentration factor and the time for which particles are trapped inside the halo by the Galactic magnetic field. This time is rather uncertain and also energy-dependent. At 100 MeV one expects roughly $\zeta \sim 10^4$ for protons or antiprotons.

MacGibbon and Carr (1991) originally calculated the PBH density required to explain the interstellar antiproton flux at 1 GeV and found a value somewhat larger than the density associated with the EGB limit. Later Maki et al. (1996) tried to fit the antiproton flux measured by the BESS balloon experiment below 0.5 GeV by using Monte Carlo simulations of cosmic ray propagation. They found that the local PBH-produced antiproton flux is mainly due to PBHs exploding within a few kpc and inferred a limit on the local PBH explosion rate of $\mathscr{R} < 0.017\,\mathrm{pc}^{-3}\,\mathrm{yr}^{-1}$. An attempt to fit more recent data leads to $\beta'(M_*) \approx 5 \times 10^{-28}$ (Barrau et al. 2003). This is well below the EGB limit, which suggests that the PBHs required to explain the EGB would overproduce antiprotons. However, these results all depend upon the PBH distribution and a different clustering assumption would lead to a different constraint. The value of ζ is also very uncertain, so this limit is not shown in Fig. 4.9. However, the figure does show the (firmer but weaker) limit associated with *extragalactic* antiprotons.

4.7.3 PBH Explosions

The EGB limit implies that the PBH explosion rate $\mathcal{R}$ could be at most $10^{-6}\,pc^{-3}\,yr^{-1}$ if the PBHs are uniformly distributed or $10\,pc^{-3}\,yr^{-1}$ if they are clustered inside the Galactic halo (Porter and Weekes 1979). The latter figure might be compared to the Galactic γ-ray limit of $0.06\,pc^{-3}\,yr^{-1}$ (Lehoucq et al. 2009) and the antiproton limit of $0.02\,pc^{-3}\,yr^{-1}$ (Maki et al. 1996). We now compare these limits to the direct observational constraints on the explosion rate.

In the Standard Model of particle physics, where the number of elementary particle species never exceeds around 100, it has been appreciated for a long time that the chances of detecting the final explosive phase of PBH evaporations are poor (Semikoz 1994). However, the physics of the QCD phase transition is still uncertain and the prospects of detecting explosions would be improved in less conventional particle physics models. For example, in a Hagedorn-type picture, where the number of particle species exponentiates at the quark-hadron temperature, the upper limit on $\mathcal{R}$ is reduced to $0.05\,pc^{-3}\,yr^{-1}$ which is comparable to the antiproton limit.

Even without the Hagedorn effect, something dramatic may occur at the QCD temperature since the number of species being emitted increases dramatically. For this reason, Cline and colleagues (e.g. Cline and Hong 1992) have long argued that the formation of a fireball at the QCD temperature could explain some of the short period γ-ray bursts (i.e. those with duration less than 100 ms). They claim that the BATSE data contains 42 candidates of this kind and the fact that their distribution matches the spiral arms suggests that they are Galactic. They also identify a class of short-period hard-spectrum KONUS bursts and eight Swift candidates with exploding PBHs. Overall they claim that the BATSE, KONUS and Swift data correspond to a 4.5σ effect and that several events exhibit the time structure expected of PBH evaporations (Cline and Otwinowski 2009).

We have seen that MacGibbon et al. have contested the claim that evaporating black holes form QCD photospheres but they accept that a photosphere might form for a short period around the QCD temperature, so perhaps the best strategy is to accept that our understanding of such effects is incomplete and focus on the empirical aspects of the γ-ray burst observations.

At much higher energies, several groups have looked for 1–100 TeV photons from PBH explosions using cosmic ray detectors. However, the constraints are again strongly dependent on the theoretical model (Petkov et al. 2008). In the Standard Model the upper limits on the explosion rate are $5 \times 10^8\,pc^{-3}\,yr^{-1}$ from the CYGNUS array, $8 \times 10^6\,pc^{-3}\,yr^{-1}$ from the Tibet array, $1 \times 10^6\,pc^{-3}\,yr^{-1}$ from the Whipple Cerenkov telescope, and $8 \times 10^8\,pc^{-3}\,yr^{-1}$ from the Andyrchy array. Full references can be found in CKSY. These limits are far weaker than the ones associated with the

EGB at 100 MeV and they would be even weaker in the QED photosphere model since there are then fewer TeV particles.

4.7.4 CMB Anisotropy

Photons emitted from PBHs sufficiently early will be completely thermalised and merely contribute to the photon-to-baryon ratio. The requirement that this does not exceed the observed ratio of around 10^9 leads to a limit

$$\beta'(M) < 10^9 \left(\frac{M}{M_P}\right)^{-1} \approx 10^{-5} \left(\frac{M}{10^9\,\text{g}}\right)^{-1} \quad (M < 10^9\,\text{g}), \qquad (4.41)$$

so only PBHs below 10^4 g could generate *all* of the CMB. This limit is shown in Fig. 4.1. Photons from PBHs in the range 10^{11} g $< M < 10^{13}$ g, although partially thermalised, will produce noticeable distortions in the CMB spectrum. Those emitted after the freeze-out of double-Compton scattering ($t \gtrsim 7 \times 10^6$ s), corresponding to $M > 10^{11}$g, induce a μ-distortion, while those emitted after the freeze-out of the single-Compton scattering ($t \gtrsim 3 \times 10^9$ s), corresponding to $M > 10^{12}$g, induce a y-distortion. The precise form of these constraints (Tashiro and Sugiyama 2008) are shown in Fig. 4.9 but they are weaker than the BBN constraint.

Another constraint on PBHs evaporating after the time of recombination is associated with the damping of small-scale CMB anisotropies. CKSY obtain the limit

$$\beta'(M) < 3 \times 10^{-30} \left(\frac{f_H}{0.1}\right)^{-1} \left(\frac{M}{10^{13}\,\text{g}}\right)^{3.1} \quad (2.5 \times 10^{13}\,\text{g} \lesssim M \lesssim 2.4 \times 10^{14}\,\text{g}), \qquad (4.42)$$

where $f_H \approx 0.1$ is the fraction of emission in electrons and positrons. Here the lower mass limit corresponds to black holes evaporating at recombination and the upper one to those evaporating at a redshift 6, after which the ionisation ensures the opacity is too low for the emitted electrons and positrons to heat the matter much. Equation (4.42) is stronger than all the other available limits in this mass range. Also shown in Fig. 4.9 is the potential limit from 21 cm observations (Mack and Wesley 2008).

4.7.5 Planck Mass Relics

If PBH evaporations leave stable Planck-mass relics, these might also contribute to the dark matter. This was first pointed out by MacGibbon (1987) and has subsequently been explored in the context of inflationary scenarios by numerous authors (Barrow et al. 1992; Carr et al. 1994; Green and Liddle 1997; Chen and Adler 2003; Barrau et al. 2003; Alexander and Meszaros 2007). If the relics have a mass κM_P and reheating occurs at a temperature T_R, then the requirement that they have less than the density $\Omega_{CDM} \approx 0.25$ implies (Carr et al. 1994)

$$\beta'(M) < 2 \times 10^{-28} \kappa^{-1} \left(\frac{M}{M_P}\right)^{3/2} \qquad (4.43)$$

for the mass range

$$\left(\frac{T_R}{T_P}\right)^{-2} < \frac{M}{M_P} < 10^{11} \kappa^{2/5}. \qquad (4.44)$$

This constraint is shown in Fig. 4.1. The lower mass limit arises because PBHs generated before reheating are diluted exponentially. The upper mass limit arises because PBHs larger than this dominate the total density before they evaporate, in which case the final cosmological photon-to-baryon ratio is determined by the baryon asymmetry associated with their emission. Indeed, in an extended inflationary scenario, evaporating PBHs may naturally generate the dark matter, the entropy and the baryon asymmetry of the Universe. This triple coincidence applies providing inflation ends at $t \sim 10^{-23}$ s, so that the PBHs have an initial mass $M \sim 10^6$ g, and this just corresponds to the upper limit indicated in Eq. (4.44). Note that Eq. (4.43) applies even if there is no inflationary period and it then extends all the way down to the Planck mass.

4.8 Conclusions

Although none of the effects discussed in this chapter provides positive evidence for PBHs, Fig. 4.9 illustrates that even the non-detection of PBHs allows one to infer important constraints on the early Universe. In particular, the limits on $\beta(M)$ can be used to constrain all the PBH formation mechanisms described in Sect. 4.2.1. Thus, for example, they constrain models involving inflation, a dustlike phase and the collapse of cosmic strings or domain walls. They also restrict the form of the primordial inhomogeneities (whatever their source) and their possible non-Gaussianity. Finally, we note that they constrain less conventional models, such as those involving a variable gravitational constant or extra dimensions. In the latter context, it should be stressed that the existence of large extra dimensions would have important implications for PBH formation, even though the accelerator black holes associated with TeV quantum gravity are not themselves primordial. However, we do not discuss this further here.

References

Abrams, N.E., Primack, J.R.: The New Universe and the Human Future (Yale University Press 2011)

Afshordi, N., McDonald, P., Spergel, D.N.: Ap. J. Lett. **594**, L71–L74 (2003)
Alexander, S., Meszaros, P.: hep-th/0703070 (2007)
Barrau, A., et al.: Astron. Astrophys. **398**, 403–410 (2003)
Barrau, A., Blais, D., Boudoul, G., Polarski, D.: Phys. Lett. B **551**, 218–225 (2003)
Barrow, J.D., Copeland, E.J., Liddle, A.R.: Phys. Rev. D **46**, 645–657 (1992)
Barrow, J.D.: Phys. Rev. D **46**, 3227–3230 (1992)
Barrow, J.D., Carr, B.J.: Phys. Rev. D **54**, 3920–3931 (1996)
Bean, R., Maguiejo, J.: Phys. Rev. D **66**, 063505 (2002)
Belyanin, A.A., et al.: MNRAS **283**(1996), 626 (1996)
Bernardeau, F., Kofman, L., Uzan, J.P.: Phys. Rev. D **70**, 083004 (2004)
Blais, D., Bringmann, T., Kiefer, C., Polarski, D.: Phys. Rev. D **67**, 024024 (2003)
Bond, J.R., Carr, B.J.: MNRAS **207**, 585–609 (1986)
Bugaev, E.V., Konishchev, K.V.: Phys. Rev. D **66**, 084004 (2002)
Bugaev, E.: Phys. Atom. Nuc. **66**, 476–480 (2003)
Bullock, J.S., Primack, J.R.: Phys. Rev. D **55**, 7423–7439 (1997)
Byrnes, C.T., Copeland. E.J., Green, A.M.: Phys. Rev. D **86**, 043512 (2012)
Caldwell, R., Casper, P.: Phys. Rev. D **53**, 3002–3010 (1996)
Carr, B.J., Hawking, S.W.: MNRAS **168**, 399–415 (1974)
Carr, B.J.: Ap. J. **201**, 1–19 (1975)
Carr, B.J.: Ap. J. **206**, 8–25 (1976)
Carr, B.J.: Astron. Astr. **56**, 377–383 (1977)
Carr, B., Rees, M.: Mon. Not. R. Astron. Soc. **206**, 801–818 (1984)
Carr, B.J., Lidsey, J.E.: Phys. Rev. D **48**, 543–553 (1993)
Carr, B.J., Gilbert, J.H., Lidsey, J.E.: Phys. Rev. D **50**, 4853–4867 (1994)
Carr, B.J., MacGibbon, J.H.: Phys. Rep. **307**, 141–154 (1998)
Carr, B.J., Kohri, K., Sendouda, Y., Yokoyama, J.: Phys. Rev. D **81**, 104019 (2010)
Chen, P., Adler, R.J.: Nucl. Phys. B **124**, 103–106 (2003)
Choptuik, M.W.: Phys. Rev. Lett. **70**, 9–12 (1993)
Cline, D., Otwinowski, S.: arXiv:0908.1352 (2009)
Cline, D.B., Hong, W.: Ap. J. Lett. **401**, L57–L60 (1992)
Cline, J., Mostoslavsky, M., Servant, G.: Phys. Rev. D **59**, 063009 (1999)
Crawford, M., Schramm, D.N.: Nature **298**, 538–540 (1982)
Daghigh, R., Kapusta, J.: Phys. Rev. D **73**, 124024 (2006)
Dokuchaev, V.I., Eroshenko, Y.N., Rubin, S.G.: astro-ph/0412418 (2004)
Dolgov, A.D., Naselsky, P.D., Novikov, I.D.: Grav. Cos. **11**, 99–104 (2000)
Duchting, N.: Phys. Rev. D **70**, 064015 (2004)
Freese, K., Price, R., Schramm, D.N.: Astrophys. J. **275**, 405–412 (1983)
Garriga, J., Sakellariadou, M.: Phys. Rev. D **48**, 2502–2515 (1993)
Gilbert, J.: Phys. Rev. D **52**, 5486–5497 (1995)
Green, A.M., Liddle, A.R.: Phys. Rev. D **56**, 6166–6174 (1997)
Green, A.M., Liddle, A.R.: Phys. Rev. D **60**, 063509 (1999)
Green, A.M., Malik, K.A.: Phys. Rev. D **64**, 021301 (2001)
Green, A.M., Liddle, A.R., Malik, K.A., Sasaki, M.: Phys. Rev. D **70**, 041502 (2004)
Halzen, F., Zas, E., MacGibbon, J., Weekes, T.C.: Nature **298**, 538–815 (1991)
Harada, T., Carr, B.J., Goymer, C.A.: Phys. Rev. D **66**, 104023 (2002)
Harada, T., Carr, B.J.: Phys. Rev. D **71**, 104009 (2005)
Hawking, S.W.: MNRAS **152**, 75–78 (1971)
Hawking, S.W.: Nature **248**, 30–31 (1974)
Hawking, S.W.: Comm. Math. Phys. **43**, 199–220 (1975)
Hawking, S.W., Moss, I., Stewart, J.: Phys. Rev. D **26**, 2681–2693 (1982)
Hawking, S.W.: Phys. Lett. B **231**, 237–239 (1989)
He, P., Fang, L.Z.: Ap. J. Lett. **568**, L1–L4 (2002)
Heckler, A.: Phys. Rev. D **55**, 480–488 (1997)

Heckler, A.: Phys. Rev. Lett. **78**, 3430–3433 (1997)

Hidalgo, J.C.: arXiv:0708.3875 (2007)

Hodges, H.M., Blumenthal, G.R.: Phys. Rev. D **42**, 3329–3333 (1990)

Inoue, K.T., Tanaka, T.: Phys. Rev. Lett. **91**, 021101 (2003)

Ioka, K., Tanaka, T., Nakamura, T., et al.: Phys. Rev. D **60**, 083512 (1999)

Ivanov, P., Naselsky, P., Novikov, I.: Phys. Rev. D **50**, 7173–7178 (1994)

Ivanov, P.: Phys. Rev. D **57**, 7145 (1998)

Jedamzik, K.: Phys. Rev. D **55**, 5871–5875 (1997)

Jedamzik, K., Niemeyer, J.: Phys. Rev. D **59**, 124014 (1999)

Josan, A., Green, A., Malik, K.: Phys. Rev. D **79**, 103520 (2008)

Kapusta, J.: Phys. Rev. Lett. **86**, 1670–1673 (2001)

Khlopov, M., Barrau, A., Grain, J.: Class. Quant. Grav. **23**, 1875–1882 (2006)

Kiraly, P., et al.: Nature **293**, 120–122 (1981)

Kodama, H., Sato, K., Sasaki, M., Maeda, K.: Prog. Theor. Phys. **66**, 2052 (1981)

Kohri, K., Yokoyama, J.: Phys. Rev. D **61**, 023501 (1999)

Lake, M., Thomas, S., Ward, J.: J. High Energy Phys. **12**, 033 (2009)

Lehoucq, R., Casse, M., Casandjan, J., Grenier, I.: Astron. Astrophys. **502**, 37–43 (2009)

Lemoine, M.: Phys. Rev. D **68**, 103503 (2003)

Lindley, D.: Mon. Not. Roy. Astron. Soc. **193**, 593–601 (1980)

Lyth, D.H., Malik, K., Sasaki, M., Zaballa, I.: JCAP **0601**, 011 (2006)

MacGibbon, J.H.: Nature **329**, 308–309 (1987)

MacGibbon, J.H., Webber, B.R.: Phys. Rev. D **41**, 3052–3079 (1990)

MacGibbon, J.H., Carr, B.J.: Ap. J. **371**, 447–469 (1991)

MacGibbon, J.H.: Phys. Rev. D **44**, 376–392 (1991)

MacGibbon, J.H., Brandenberger, R.H., Wichoski, U.F.: Phys. Rev. D **57**, 2158–2165 (1998)

MacGibbon, J.H., Carr, B.J., Page, D.N.: Phys. Rev. D **78**, 0709–2380 (2008)

Mack, K., Wesley, DH.: arXiv:0805.1531 (2008)

Maki, K., Mitsui, T., Orito, S.: Phys. Rev. Lett. **76**, 3474 (1996)

Matsuda, T.: JEHP **0604**, 017 (2006)

Meszaros, P.: Astron. Astrophys. **38**, 5–13 (1975)

Miyama, S., Sato, K.: Prog. Theor. Phys. **59**, 1012 (1978)

Moss, I.G.: Phys. Rev. D **50**, 676–681 (1994)

Musco, I., Miller, J.C., Rezzolla, L.: Class. Quant. Grav. **22**, 1405–1424 (2005)

Nadezhin, D.K., Novikov, I.D., Polnarev, A.G.: Sov. Astron. **22**, 129–138 (1978)

Nakamura, T., Sasaki, M., Tanaka, T., Thorne, K.: Ap. J. Lett. **487**, L139–L142 (1997)

Niemeyer, J., Jedamzik, K.: Phys. Rev. Lett. **80**, 5481–5484 (1998)

Niemeyer, J., Jedamzik, K.: Phys. Rev. D **59**, 124013 (1999)

Novikov, I.D., Polnarev, A.G., Starobinsky, A.A., Zeldovich, Y.B.: Astron. Astrophys. **80**, 104–109 (1979)

Okele, P., Rees, M.: Astron. Astrophys. **81**, 263–264 (1980)

Page, D.N., Hawking, S.W.: Ap. J. **206**, 1–7 (1976)

Petkov, V., Bugaev, E., Klimai, P., Smirnov, D.: JETP Lett. **87**, 1–3 (2008)

Polnarev, A.G., Khlopov, M.: Sov. Phys. Usp. **28**, 213–232 (1985)

Polnarev, A.G., Zemboricz, R.: Phys. Rev. D **43**, 1106–1109 (1988)

Porter, N.A., Weekes, T.C.: Nature **277**, 199 (1979)

Rees, M.: Nature **266**, 333–334 (1977)

Ricotti, M., Ostriker, J., Mack, K.: Astrophys. J. **880**, 829–845 (2008)

Rubin, S.G., Yu, M.K., Sakharov, A.S.: Grav. Cos. **6**, 51–58 (2001)

Saito, R., Yokoyama, J.: Phys. Rev. Lett. **102**, 161101 (2009)

Semikoz, D.V.: Ap. J. **436**, 254–256 (1994)

Sendouda, Y., Kohri, K., Nagataki, S., Sato, K.: Phys. Rev. D **71**, 063512 (2005)

Shibata, M., Sasaki, M.: Phys. Rev. D **60**, 084002 (1999)

Suyama, T., Tanaka, T., Bassett, B., Kudoh, H.: Phys. Rev. D **71**, 063507 (2005)

Tashiro, H., Sugiyama, N.: Phys. Rev. D **78**, 023004 (2008)
Vainer, B., Nasel'skii, P.: Sov. Astron. **22**, 138–140 (1978)
Vainer, B., Dryzhakova, O., Naselskii, P.: Sov. Astron. Lett. **4**, 185–187 (1978)
Wright, E.L.: Ap. J. **459**, 487–490 (1996)
Yokoyama, J.: Astron. Astrophys. **318**, 673–679 (1997)
Yokoyama, J.: Phys. Rev. D **58**, 107502 (1998)
Yokoyama, J.: Phys. Rev. D **58**, 083510 (1998)
Yokoyama, J.: Prog. Theor. Phys. Supp. **136**, 338–352 (1999)
Yu, K.M., Polnarev, A.G.: Phys. Lett. B **97**, 383–387 (1980)
Yu, M.K., Malomed, B., Zeldovich, Y.B.: Mon. Not. Roy. Astron. Soc. **215**, 153–156 (1985)
Yu, K.M., Konoplich, R.V., Rubin, S.G., Sakharov, A.S.: Grav. Cos. **6**, 153–156 (2000)
Yu, K.M., Rubin, S.G., Sakharov, A.S.: J. Astropart. Phys. **23**, 265–277 (2005)
Zeldovich, Y.B.: Sov. Astron. A. J. **10**, 602–603 (1967)
Zeldovich, Y.B., Starobinski, A.: Sov. J. Exp. Theor. Phys. Lett. **24**, 571–573 (1976)
Zeldovich, Y.B., Starobinski, A., Khlopov, M., Chechetkin, V.: Sov. Astron. Lett. **3**, 110–112 (1977)

Chapter 5
Black Hole Formation in High Energy Particle Collisions

Abstract In this chapter we describe how black holes form in the high energy collision of two particles. The challenge lies in establishing that a closed trapped surface forms during that collision. Remarkably, it is possible to construct such a surface analytically for a $3 + 1$ dimensional space-time. This construction can be extended into the semi-classical regime.

5.1 Introduction

Assessing whether a region of space-time will evolve into a black hole is in general a challenging task. However, if the matter distribution has enough symmetries, this task can be easy. For example, it is straightforward to solve explicitly Einstein's equations for the gravitational field outside a spherically symmetrical object, such as a star or a planet. The solution is the famous Schwarzschild solution (2.2). It is easy to check whether an object is a black hole: one just needs to know whether its radius is smaller than its Schwarzschild radius r_s (see Sect. 2.2.1).

This exercise is more difficult in situations with fewer symmetries. A very asymmetrical case is that of two particles in a head-on collision. Determining whether a black hole will form in such a collision is important for models with low-scale quantum gravity, which we shall discuss in Chap. 6.

Motivated by the condition for a spherically symmetrical object to be a black hole if its radius is smaller that its Schwarzschild radius, Thorne proposed a criterion, the hoop conjecture, for a region of space-time to evolve into a black hole Thorne (1972). The hoop conjecture states that a region of space-time which is imploding will form a black hole if a circular hoop of circumference $C = 2\pi r_s$ can be placed around the object and rotated. To motivate this conjecture, which has subsequently been probed by numerous numerical studies, Thorne studied the effect of gravity on objects of different shapes and concluded that they need to be compressed in all three spatial directions before gravity leads to the formation of a black hole.

X. Calmet et al., *Quantum Black Holes*, SpringerBriefs in Physics,
DOI: 10.1007/978-3-642-38939-9_5, © The Author(s) 2014

The hoop conjecture is the basis for the geometrical cross-section formula for black hole production in the collision of two particles at very high energy

$$\sigma = \pi r_s^2. \tag{5.1}$$

In this case the Schwarzschild radius is related to the centre of mass energy $\sqrt{s}$ by $r_s = 2G\sqrt{s}/c^2$, i.e. one assumes that all the energy of the colliding particles is absorbed by the black hole. The mass of the black hole formed is determined by the centre of mass energy.

Determining whether two particles in a high energy collision will form a black hole is difficult. We shall now introduce a condition for the formation of a black hole, which has been used in Penrose (unpublished), D'Eath and Payne (1992), Eardley and Giddings (2002). To establish gravitational collapse, one needs to prove the existence of a closed trapped surface defined as follows. Imagine that at some instant, a sphere S emits a flash of light, see Fig. 5.1. At a later time, the light from a particular point P forms a sphere F around P. We construct the ingoing and outgoing wavefronts from the envelopes S_1 and S_2 of the light from all the points of S. If the areas of both S_1 and S_2 are less than that of S, then S is a closed trapped surface. Clearly something very strange has happened to the space-time. When a closed trapped surface is formed, the Hawking-Penrose theorem Hawking and Penrose (1970) implies that there is a singularity in the future evolution of Einstein's equations. The Hawking-Penrose theorem

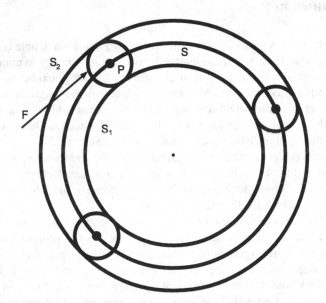

Fig. 5.1 A closed trapped surface: at some instant, a sphere S emits a flash of light. At a later time, the light from a point P forms a sphere F around P, and the envelopes S_1 and S_2 form the ingoing and outgoing wavefronts respectively. If the areas of both S_1 and S_2 are less than that of S, then S is a closed trapped surface

assumes that the energy momentum tensor satisfies the weak energy condition. The weak energy condition states that for every future pointing time-like vector field **X**, the matter density observed by the corresponding observers is always non-negative: $\rho = T_{ab}X^a X^b \geq 0$.

Furthermore one needs to assume that there are no naked singularities. One expects physical laws to break down at space-time singularities, so physics might lose its predictive power there. Penrose (1969, 2002) cosmic censorship conjecture says that singularities cannot be observed by an outside observer. In other words, all naked singularities are hidden from an observer at infinity by an event horizon and singularities in the evolution of Einstein's equations imply black hole formation. The Hawking-Penrose theorem together with the cosmic censorship conjecture leads to the conclusion that if a closed trapped surface is formed, then a black hole is formed as well.

We shall now review the current state of the art in our understanding of the formation of black holes in the collision of two particles at very high energy. We shall distinguish three cases: classical black holes with mass much larger than the Planck scale, semi-classical black holes with masses a few times larger than the Planck scale and quantum black holes with masses close or equal to the Planck scale.

5.2 Classical Black Holes

We now sketch the derivation of a closed trapped surface given by Eardley and Giddings (2002), using their notation. The construction involves studying the collision of two particles. Each particle is modeled by the Aichelburg-Sexl metric Aichelburg and Sexl (1971), which is obtained by considering the Schwarzschild solution in the limit of large boost and small mass, with fixed total energy. Each particle carries an energy $\mu = \sqrt{s}/2$ where $\sqrt{s}$ is the centre of mass energy. In d-dimensions, the result for a particle moving in the $+z$ direction is the metric

$$ds^2 = -d\bar{u}\,d\bar{v} + d\bar{x}^{i2} + \Phi(\bar{\rho})\delta(\bar{u})d\bar{u}^2, \qquad (5.2)$$

where Φ depends only on the transverse radius $\bar{\rho} = \sqrt{\bar{x}^i \bar{x}_i}$ and takes the form

$$\Phi = \begin{cases} -8G\mu \ln(\bar{\rho}), & d = 4, \\ \frac{16\pi G\mu}{\Omega_{d-3}(d-4)\bar{\rho}^{d-4}}, & d > 4, \end{cases} \qquad (5.3)$$

where Ω_{d-2} is the volume of the unit $(d-2)$-sphere. Note that Φ satisfies Poisson's equation

$$\nabla^2 \Phi = -16\pi G\mu\delta^{d-2}(\bar{x}^i) \qquad (5.4)$$

in the transverse dimensions, where ∇ is the $(d-2)$-dimensional flat-space derivative in the $(\bar{x}^i)$.

The Aichelburg-Sexl metric is manifestly flat except in the null plane $\bar{u} = 0$ of the shockwave. Eardley and Giddings then consider an identical shockwave traveling along $\bar{v} = 0$ in the $-z$ direction. Remarkably, by causality, these shockwaves will not be able to influence each other until the shockwaves collide. This means that one can superpose two solutions of the form (5.2) to give the exact geometry outside the future light cone of the collision of the shockwaves. Furthermore, there is thus no need to speculate on any details of quantum gravity. The construction however only applies to classical black holes with masses much larger than the Planck scale.

It is useful to introduce a new coordinate system defined by

$$\bar{u} = u,$$

$$\bar{v} = v + \Phi(x)\theta(u) + \frac{u\theta(u)(\nabla\Phi(x))^2}{4},$$

$$\bar{x}^i = x^i + \frac{u}{2}\nabla_i\Phi(x)\theta(u), \tag{5.5}$$

where θ is the Heaviside step function. In this coordinates geodesics and their tangents are continuous across the shockwave at $u = 0$ and the metric of the combined shockwaves becomes

$$ds^2 = -du\,dv + \left[H_{ik}^{(1)} H_{jk}^{(1)} + H_{ik}^{(2)} H_{jk}^{(2)} - \delta_{ij} \right] dx^i dx^j \tag{5.6}$$

where

$$H_{ij}^{(1)} = \delta_{ij} + \frac{1}{2}\nabla_i\nabla_j\Phi(\mathbf{x} - \mathbf{x}_1)\,u\theta(u) \tag{5.7}$$

$$H_{ij}^{(2)} = \delta_{ij} + \frac{1}{2}\nabla_i\nabla_j\Phi(\mathbf{x} - \mathbf{x}_2)\,v\theta(v) \tag{5.8}$$

with Φ given by Eq. (5.3),

$$\mathbf{x}_1 = (+b/2, 0, \dots, 0), \quad \mathbf{x}_2 = (-b/2, 0, \dots, 0), \tag{5.9}$$

and $\mathbf{x} \equiv (x^i)$ is in the transverse flat $(d-2)$-space.

For the case of $d = 4$ and zero impact parameter $(b = 0)$, the trapped surface is in the union of the two shockwaves. It consists of two flat disks with radii ρ_c at

$$\bar{t} = -4G\mu\ln\rho_c, \quad \bar{z} = \pm 4G\mu\ln\rho_c. \tag{5.10}$$

Matching their normals across the boundary, which lies in the collision surface $u = v = 0$, one finds $\rho_c = 4G\mu = r_h$. If one repeats this construction for $d > 4$ and $b = 0$, one obtains

$$\rho_c = \left(\frac{8\pi G\mu}{\Omega_{d-3}}\right)^{1/(d-3)}. \qquad (5.11)$$

Remarkably, this construction can be generalized to the case of a non-zero impact parameter. In the case $d = 4$, Eardley and Giddings were able to construct a closed trapped surface analytically while in the case $d > 4$ one has to rely on numerical solutions Eardley and Giddings (2002), Yoshino and Nambu (2003), Yoshino and Rychkov (2005). The heart of the problem is finding a closed trapped surface $\mathscr{S}$ lying in the union of the hypersurfaces describing the incoming shockwaves. These null hypersurfaces are defined by $v \leq 0 = u$ and $u \leq 0 = v$ and intersect in the $(d-2)$-dimensional surface $u = 0 = v$. This latter surface intersects the closed trapped surface $\mathscr{S}$ in a closed $(d-3)$-dimensional surface $\mathscr{C}$. The challenge therefore is determining $\mathscr{C}$.

In the first incoming null surface $v \leq 0 = u$, we define $\mathscr{S}$ by

$$v = -\Psi_1(\mathbf{x}) \quad \text{with } \Psi_1 > 0 \text{ inside } \mathscr{C} \text{ and } \Psi_1 = 0 \text{ on } \mathscr{C}. \qquad (5.12)$$

For $\mathscr{S}$ to be a closed trapped surface, we require neighbouring light rays normal to $\mathscr{S}$ to move towards each other, which, in technical language, means that the outer null normals have zero convergence. For $v < 0$ this happens as long as

$$\nabla^2(\Psi_1 - \Phi_1) = 0 \quad \text{inside } \mathscr{C}. \qquad (5.13)$$

Similarly, in the second incoming null surface $u \leq 0 = v$, defining $\mathscr{S}$ by

$$u = -\Psi_2(\mathbf{x}) \quad \text{with } \Psi_2 > 0 \text{ inside } \mathscr{C} \text{ and } \Psi_2 = 0 \text{ on } \mathscr{C}, \qquad (5.14)$$

the outer null normals have zero convergence for $u < 0$ providing

$$\nabla^2(\Psi_2 - \Phi_2) = 0 \quad \text{inside } \mathscr{C}. \qquad (5.15)$$

The final condition is that the outer null normal to $\mathscr{S}$ must be continuous across $u = 0 = v$. A necessary and sufficient condition for this continuity is that

$$\nabla\Psi_1 \cdot \nabla\Psi_2 = 4 \quad \text{on } \mathscr{C} ; \qquad (5.16)$$

since Ψ_1 and Ψ_2 vanish on $\mathscr{C}$, $\nabla_i\Psi_\alpha$ is normal to $\mathscr{C}$.

Note that (5.4, 5.13) imply that Ψ_α satisfies Poisson's equation with sources at $\mathbf{x}_\alpha$. One can define the rescaled functions

$$g(\mathbf{x}, \mathbf{x}_\alpha; \mathscr{C}) = \frac{\Omega_{d-3}}{16\pi G\mu}\Psi_\alpha \qquad (5.17)$$

satisfying

$$\nabla_x^2 g(\mathbf{x}, \mathbf{x}_\alpha; \mathscr{C}) = -\Omega_{d-3} \delta^{d-2}(\mathbf{x} - \mathbf{x}_\alpha), \qquad (5.18)$$

$$g(\mathbf{x}, \mathbf{x}_\alpha; \mathscr{C}) = 0 \quad \text{for } \mathbf{x} \text{ on } \mathscr{C}. \qquad (5.19)$$

Therefore the $g(\mathbf{x}, \mathbf{x}_\alpha; \mathscr{C})$ are Dirichlet Green's functions for sources at $\mathbf{x}_1, \mathbf{x}_2$ with boundary $\mathscr{C}$. The above construction has mapped the problem of finding the trapped surface onto an equivalent simple mathematical problem, which is given as follows. Let $\mathbf{x}_1$ and $\mathbf{x}_2$ be two points in $(d-2)$-dimensional Euclidean space, and let $B > 0$ be a constant. If $g(\mathbf{x}, \mathbf{x}_\alpha; \mathscr{C})$ are the Dirichlet Green's functions satisfying (5.18, 5.19), then the problem is to find a closed $(d-3)$-surface $\mathscr{C}$ enclosing the points $\mathbf{x}_1$ and $\mathbf{x}_2$ and having the following property:

$$\nabla_{\mathbf{x}} g(\mathbf{x}, \mathbf{x}_1; \mathscr{C}) \cdot \nabla_{\mathbf{x}} g(\mathbf{x}, \mathbf{x}_2; \mathscr{C}) = B^2 \qquad (5.20)$$

for all points $\mathbf{x}$ on $\mathscr{C}$.

In four dimensions, one has

$$\Psi_1 = 8G\mu g(\mathbf{x}, \mathbf{x}_1; \mathscr{C}), \qquad \Psi_2 = 8G\mu g(\mathbf{x}, \mathbf{x}_2; \mathscr{C}) \qquad \text{and} \qquad B = \frac{1}{4G\mu}.$$

Using the following definitions

$$\mathbf{x}_1 = \left(\frac{2G\mu(1-a^2)}{a} \ln\left(\frac{1+a^2}{1-a^2}\right), 0 \right) = -\mathbf{x}_2,$$

$$b(a) = \frac{4G\mu(1-a^2)}{a} \ln\left(\frac{1+a^2}{1-a^2}\right), \qquad (5.21)$$

where the parameter a is such that $0 \leq a < 1$, Eardley and Giddings construct a closed trapped surface $\mathscr{S}$ for any value of the impact parameter $b(a)$. The area of $\mathscr{S}$ is found to be

$$\text{Area}(\mathscr{S}) = 16\pi(G\mu)^2 \frac{(1-a^2)^2}{a^2} \ln\left(\frac{1+a^2}{1-a^2}\right). \qquad (5.22)$$

An apparent horizon is defined as the outermost trapped surface in space-time, and always lies inside the event horizon of a black hole. Since we have a closed trapped surface $\mathscr{S}$, either $\mathscr{S}$ is itself the apparent horizon or there is a apparent horizon outside $\mathscr{S}$. Therefore, Area$(\mathscr{S})$ gives a lower bound gives a lower bound on the area of the apparent horizon because $\mathscr{S}$ is convex and the 2-metric is Euclidean. As the apparent horizon area is less than or equal to the Schwarzschild horizon area $4\pi r_s^2$, there is a lower bound on the mass of the final black hole:

$$M_{\text{final bh}} > 2\mu \frac{1-a^2}{2a} \sqrt{\ln\left(\frac{1+a^2}{1-a^2}\right)}. \qquad (5.23)$$

Furthermore, the fraction of total energy $2\mu = \sqrt{s}$ emitted as gravitational radiation is bounded from above,

$$\frac{E_{\text{grav rad}}}{2\mu} < 1 - \frac{1-a^2}{2a}\sqrt{\ln\left(\frac{1+a^2}{1-a^2}\right)}. \tag{5.24}$$

As a matter of fact, $E_{\text{grav rad}}$ may be significantly smaller because the final black hole is expected to be rotating, unless the impact parameter b is zero, and thus a substantial proportion of its energy should be in the form of rotational energy.

The function $b(a)$, for a given μ, reaches a maximum value of

$$b_{\text{max}} \approx 3.219 G\mu \tag{5.25}$$

at

$$a_{\text{max}} \approx 0.6153. \tag{5.26}$$

This is the largest value of the impact parameter for which one can show, by this method, that a black hole is produced in the collision of two particles. The corresponding lower limit on the cross-section is

$$\sigma \geq \pi b_{\text{max}}^2 \approx 32.552 (G\mu)^2. \tag{5.27}$$

The production cross-section for black holes in the collision of two particles predicted by the hoop conjecture is $\sigma = \pi r_s^2$. The lower limit obtained by Eardley and Giddings is about 65 % of this estimate. Thus not all of the energy of the colliding particles is transferred to the black hole. Some will be radiated away (see Sect. 3.3) and is not available for gravitational collapse. Using (5.23), Eardley and Giddings find a lower bound on the final black hole mass ranging from $0.71\sqrt{s}$ for $b = 0$ to $0.45\sqrt{s}$ for $b = b_{\text{max}}$.

Thus far we have been dealing with the classical production of black holes. One may worry that when trying to extend this construction at the quantum mechanical level, the production could be suppressed exponentially, as is the case for tunneling processes which are only allowed quantum mechanically and not classically. However, it is important to realize that the construction due to Eardley and Giddings is purely classical. This implies that there is no Euclidean suppression in the formation of black holes in the collisions of two particles at the quantum level. The reason is that there are classical trajectories with two particle initial conditions which evolve into black holes. The process is not classically forbidden, as demonstrated by Eardley and Giddings. There is thus no tunneling factor. One might also worry that since black holes produce a thermal spectrum of particles during their evaporation in the form of Hawking radiation (see Chap. 3) rather than a few highly energetic particles, the time-reversal of the production process, and hence the production process itself, by CPT (i.e. time reversal invariance) must have very low probability. However, the time-reversed classical solutions exhibit a very energetic wave of gravitational radiation colliding with the time-reversed black hole to produce the two particle state. This is the energy which escapes the black hole in the formation process. The process is not thermal. It involves very special initial and final conditions. These two points have been emphasized in Hsu (2003). Note that the particles are assumed to be point-like in Eardley and Giddings' construction. This construction can be extended to the case of finite size particles Giddings and Rychkov (2004), Kohlprath and Veneziano (2002).

This beautiful construction is purely classical and most remarkably it does not require any assumption on the nature of quantum gravity. It holds, as we have emphasized already, in the limit where the centre of mass energy $\sqrt{s}$ is much larger than the Planck scale and the black hole mass. Despite its elegance, this construction is of limited practical importance for realistic collisions where the centre of mass energy would certainly not be much above the Planck scale. Fortunately, a path integral formulation of the construction has been proposed by Hsu (2003). This work extends the validity of Eardley and Giddings' result into the semi-classical regime.

5.3 Semi-classical Black Holes

Semi-classical black holes are black holes with masses somewhat larger than the Planck scale. In general relativity, the Planck mass M_P is approximately 2×10^{-8} kg. The precise factor is dependent on the model of gravity and is actually poorly known. An educated guess is that the semi-classical regime starts at black hole masses five to twenty times larger than the Planck scale.

We shall now describe the extension of the construction of Eardley and Giddings into the semi-classical regime. Although the initial colliding particles are dressed by strong gravitational fields, from a quantum field theoretical point of view, the initial state will be described by a two-particle state. A semi-classical black hole is a thermal object and thus expected to decay via Hawking radiation (see Chap. 3) to a large number of particles. We thus need to describe amplitudes involving two particles with high momenta in the initial states going to many particles with low momenta in the final state via a black hole $\mathscr{A} = \langle p_1 p_2 | \text{semi} - \text{classical BH} | \sum_i p_i \rangle$.

In Hsu (2003), Hsu proposed using an effective action Gould et al. (1995) designed to treat such processes. While general covariance does not permit a unique time slicing, the gravitational action $S_g = \int d^4 x \sqrt{-g} \, R$ is still well-defined. The difficulty in defining a path integral is specifying over which field configurations one should sum. One is interested in describing semi-classical black holes. How does one include them in the path integral? Hsu emphasizes that the notion of an S-matrix is appropriate if one considers asymptotically flat space-times in the far past and future, on top of additional excitations. Black holes are assumed to be excitations, i.e. particle states, and the Hilbert space must thus be extended to include quantum states representing black holes. One can define semi-classical black hole states pragmatically as those with a strong overlap with the trajectories corresponding to classical black holes. In a classical black hole solution, excess energy is radiated away, see Chap. 3, in the form of Hawking radiation by late times and the exterior metric can be classified by a limited number of quantum numbers such as mass, charge and angular momentum. This is a consequence of the famous no hair theorem Misner et al. (1973). A minimal formulation thus involves a Hilbert space of black holes classified by their exterior metric at future null infinity (I^+, see for example Fig. 2.2).

For the semi-classical approximation to be useful, one needs to show that there are no large quantum corrections. In general relativity there is no small dimensionless parameter. However, expanding about a background configuration yields interactions which are suppressed by the background curvature in Planck units. Classical solutions describing the ordinary gravitational collapse of many soft particles, for example the collapse of a large star or dust ball, can produce black holes without regions of large curvature. Hsu's formalism applies directly to such solutions, resulting in a semi-classical amplitude without large quantum corrections.

In the construction of Eardley and Giddings, regions of large curvature can arise quite early in the evolution even if the black hole produced is large, $\mu \gg M_p$. If one takes the size of the colliding particles to be of order the Planck length, one finds curvatures at the shock front of order μ^2. In this case, quantum corrections might be large. Because gravity is non-renormalizable, one has to consider all possible generally covariant higher dimension operators in the Lagrangian, such as higher powers of the curvature. In large curvature backgrounds these terms may not be negligible, so the size of quantum corrections will in principle depend on unknown details of quantum gravity. Thus, this construction cannot be applied to black holes with mass close to the Planck scale. However, for the parameter range relevant to models with low scale quantum gravity to be described in Chap. 6, the semi-classical approximation is expected to be a good one. As explained by Hsu, this supposes that quantum gravitational effects are not sensitive to the size of the objects colliding.

5.4 Planckian Quantum Black Holes

Planckian quantum black holes are black holes with masses very close to the Planck scale. They are thus non-thermal objects. These black holes are the most quantum objects in quantum gravity and very little is known about their creation and decay properties. They could either be stable objects, in which case they are called remnants also called relics, or very short lived and decay immediately after having been created. If they decay, their decomposition is not expected to be described accurately by Hawking radiation as they are non-thermal objects. They are expected to decay to only a few particles. Their production cross-section is usually extrapolated from the semi-classical one. Symmetries have been used to constrain the possible decay products Calmet et al. (2008). Furthermore, while semi-classical black holes are expected to have a continuous mass spectrum, quantum black holes could have either a continuous or discrete mass spectrum Calmet et al. (2012). Planckian quantum black hole physics is the topic of on-going research.

Let us finish by emphasizing that evaporating astrophysical black holes may conceivably leave stable Planck mass relics, in which case these relics could provide another solution to the dark matter problem Carr (1994).

5.5 Conclusions

In this chapter, we discussed the current state of the art in the formation of small black holes in the collisions of two high energetic particles. We emphasized that progress has gone well beyond the hoop conjecture. A mathematical proof of gravitational collapse in such processes is known. Finally we explained that it is important to differentiate between classical, semi-classical and quantum black hole production mechanics. While in the first two cases, we can predict with confidence when a black hole will form, the latter case is still work in progress.

References

Aichelburg, P.C., Sexl, R.U.: Gen. Rel. Grav. 2, 303–312 (1971)
Calmet, X., Fragkakis, D., Gausmann, N.: Non-thermal small black holes. In: Bauer, A.J., Eiffel, D.G. (eds.) Black Holes: Evolution, Theory and Thermodynamics, pp. 165–170. Nova Science Publishers, New York (2012)
Calmet, X., Gong, W., Hsu, S.D.H.: Phys. Lett. B 668, 20–23 (2008)
Carr, B.J.: Ann. Rev. Astron. Astrophys. 32, 531–590 (1994)
D'Eath, P.D., Payne, P.N.: Phys. Rev. D 46, 658–674 (1992)
D'Eath, P.D., Payne, P.N.: Phys. Rev. D 46, 675–693 (1992)
D'Eath, P.D., Payne, P.N.: Phys. Rev. D 46, 694–701 (1992)
Eardley, D.M., Giddings, S.B.: Phys. Rev. D 66, 044011 (2002)
Giddings, S.B., Rychkov, V.S.: Phys. Rev. D 70, 104026 (2004)
Gould, T.M., Hsu, S.D.H., Poppitz, E.R.: Nucl. Phys. B 437, 83–106 (1995)
Hawking, S.W., Penrose, R.: Proc. Roy. Soc. Lond. A 314, 529–548 (1970)
Hsu, S.D.H.: Phys. Lett. B 555, 92–98 (2003)
Kohlprath, E., Veneziano, G.: JHEP 0206, 057 (2002)
Misner, C.W., Thorne, K.S., Wheeler, J.A.: Gravitation. W. H. Freeman, San Francisco (1973)
Penrose, R., Riv. Nuovo Cim. 1, 252–276 (1969)
Penrose, R.: Gen. Rel. Grav. 34, 1141–1165 (2002)
Thorne, K.S.: Nonspherical gravitational collapse: a short review. In: J.R. Klauder (ed.), Magic Without Magic, pp. 231–258. W. H. Freeman, San Francisco (1972)
Yoshino, H., Rychkov, V.S.: Phys. Rev. D 71, 104028 (2005) [Erratum-ibid. D 77, 089905 (2008)]
Yoshino, H., Nambu, Y.: Phys. Rev. D 67, 024009 (2003)

Chapter 6
Black Holes and Low Scale Quantum Gravity

Abstract In this chapter we describe models with low scale quantum gravity. While clearly speculative, these models have been very useful in demonstrating that we do not know from first principles the energy scale at which quantum gravitational effects become large. We emphasize that it is thus important to search for quantum gravitational effects, including the formation of small black holes in experiments such as the Large Hadron Collider and cosmic ray experiments.

6.1 Models of Low Scale Quantum Gravity

One of the most exciting developments in theoretical physics in the last 15 years has been the realization that we know very poorly the energy scale at which quantum gravitational interactions become important. The Planck scale is the energy scale above which the quantum fluctuations of space-time cannot be neglected. Traditionally, this energy scale was thought to be around 10^{19} GeV. This assumption was based on the construction of a number having the dimensions of mass, i.e. the Planck mass $M_p = \sqrt{\hbar c/G} = 1.2209 \times 10^{19}$ GeV/c^2 where G is Newton's constant, c is the speed of light and $\hbar$ the reduced Planck constant. We will make a distinction between the Planck mass with units of GeV/c^2 and the Planck scale which is an energy with units of GeV.

Until recently, it was assumed that the Planck scale might never be probed experimentally. However, Arkani-Hamed et al. (1998); Antoniadis et al. (1998) have shown by studying a class of models with more than four dimensions that the Planck scale could be much lower than naively expected and potentially around a few 10^3 GeV depending on the number of extra dimensions. Their work is remarkable in demonstrating that we do not know at what scale quantum gravitational effects become large. Their models motivate a search for quantum gravitational effects using Earth-based experiments such as colliders or cosmic ray detectors. Most remarkably, some of these models are compatible with all current observations in particle physics and in cosmology.

These models posit that we live in more than four dimensions. This, as such, is not revolutionary as string theorists have been pushing this idea for decades. However, in string theory, the extra dimensions are typically curled up and very tiny. The proposal of Arkani-Hamed et al. is shocking because the volume of the extra dimensions has to be comparatively large. The size of the extra dimensions could be in the millimeter range. Starting from a higher dimensional action of the form

$$S = \int d^4x \, d^{d-4}x' \sqrt{-g} \left(M_\star^{d-2} \mathcal{R} + \cdots \right) \tag{6.1}$$

one finds that the effective $3 + 1$ effective Planck scale M_P is given by

$$M_p^2 = M_\star^{d-2} V_{d-4} \tag{6.2}$$

where V_{d-4} is the volume of the extra dimensions. By taking V_{d-4} large, M_p can be made of order 10^{19} GeV while $M_\star \sim$ TeV. Clearly this comes at the cost of some strong dynamical assumptions about the geometry of space-time: why are some dimensions larger than others?

There are different realizations of this idea. We first describe the brane world model proposed by Arkani-Hamed et al. (1998); Antoniadis et al. (1998), in which model particles of the standard model are confined to a three dimensional surface, called the brane, whereas gravity can propagate everywhere both on the brane and in the extra-dimensional volume called the bulk. The number of extra dimensions is not fixed from first principles, but can only be constrained by experiments. There are bounds coming from astrophysics, searches for modifications of Newton's $1/r$ potential on Earth and collider experiments. With the Large Hadron Collider at CERN accumulating data at an impressive pace, these bounds are evolving quickly and we refer the reader to The Review of Particle Physics Beringer (2012), which is regularly updated, for up-to-date limits on the parameters of these models. A Planck scale of 1 TeV is excluded for one extra dimension by the non-observation of modifications of Newtonian gravity within the solar system. It is possible to set bounds on the Planck scale in models with two and three extra dimensions using neutron stars, supernovae physics and cosmological observations. The most stringent ones come from neutron star physics, which would be affected quite dramatically by the presence of Kaluza-Klein excitations of the graviton predicted by these models. One finds $M_P > 1700$ TeV and $M_P > 76$ TeV for $n = 2$ and 3 extra-dimensions respectively Beringer (2012). For $n \geq 4$, colliders are setting limits on the Planck scale which should be larger than a few TeV.

In the version proposed by Randall and Sundrum (RS) Randall and Sundrum (1999); Gogberashvili (2002), a five-dimensional space-time is considered with two branes. In the simplest version of Randall and Sundrum's model, the standard model particles are confined to the so-called infra-red brane while gravity propagates in the bulk as well. For this model, the bounds on the Planck scale are typically in the TeV region.

One of the main difficulties of models with large extra dimensions is proton decay. In the Randall and Sundrum model, it was proposed that leptons and quarks could propagate in the bulk in order to suppress proton decay operators Huber (2003). However, this leads to tighter bounds on the parameters of the model due to the non-observations of certain rare or forbidden decays of particles of the standard model.

More recently, it has been shown that the Planck scale could also be lower than naively expected, even if there are only four space-time dimension, if there is a large hidden sector of particles which interact only gravitationally with the standard model Calmet et al. (2008). This construction is based on the fact that Newton's constant and hence the Planck mass are scale-dependent quantities, like any other coupling constant of a quantum field theory. The actual scale μ_* at which quantum gravitational effects are large is given by

$$M(\mu_*) \sim \mu_*, \tag{6.3}$$

where $M(\mu)$ is the running Planck mass as a function of the energy scale μ. This condition implies that quantum fluctuations in space-time geometry at length scales μ_*^{-1} will be unsuppressed. The contributions of spin 0, spin 1/2 and spin 1 particles to the running of $M(\mu)$ can easily be calculated using the heat-kernel method. This regularization procedure ensures that the symmetries of the theory are preserved by the regulator. One finds Vassilevich (1995); Larsen and Wilczek (1996); Kabat (1995); Calmet et al. (2008)

$$M(\mu)^2 = M(0)^2 - \frac{\mu^2}{12\pi} N \tag{6.4}$$

where $M(0)$ is the Planck mass measured in low energy experiment. In this equation $N = N_0 + N_{1/2} - 4N_1$, where the parameters N_0, $N_{1/2}$ and N_1 are the number of scalar fields, Weyl fermions and gauge bosons in the theory respectively. This result relies only on quantum field theory in curved space-time and does not require any assumption about quantum gravity.

There are thus several models with can lead to a Planck scale which is lower than naively expected using dimensional analysis.

6.2 Microscopic Black Holes at Colliders

If the scale of quantum gravity is truly as low as a few TeV, colliders such as the Large Hadron Collider (LHC) could produce microscopic black holes Dimopoulos and Landsberg (2001); Banks and Fischler (1999); Giddings and Thomas (2002); Feng and Shapere (2002); Anchordoqui et al. (2004, 2002, 2003); Meade and Randall

(2008); Calmet et al. (2008). At the LHC, the centre-of-mass energy of the collision between the colliding protons will eventually increase up to 14 TeV. In the first few years, collisions took place at 7 and 8 TeV. Clearly this will allow us to probe a Planck mass in the few TeV region. The most massive black hole which could be produced at the LHC would at most have a mass of 14 TeV and likely less since, as we shall see, not all of the energy of the protons is available for gravitational collapse. This implies that the ratio between the black hole mass and the Planck mass will be at most around 14 in the most optimistic case, i.e. a Planck scale of 1 TeV. These black holes are thus at best semi-classical and therefore very different from their astrophysical counterparts. The production mechanism for microscopic black holes in the collision of two particles has been discussed in Chap. 5.

It is important to realize that for black hole production, as for the production of any other particle at a proton collider, what matters are collisions between the constituents, the so-called partons, of the protons namely the quarks, anti-quarks and gluons. The partons carry the fraction of the energy of the proton available to form a black hole or to produce any other particle. Furthermore, Eardley and Giddings (2002) have found that only part of the energy of the partons is available for gravitational collapse (see Chap. 5). The fraction of energy carried by two colliding partons is parametrized by v and u/v. The parton distribution function $f_i(v, Q)$ describes the probability of finding a parton of type i with energy squared sv, where s is the centre-of-mass energy squared, in the colliding proton. The energy scale Q corresponds to the momentum transfer and in black hole production processes is often associated with the Planck scale. The production cross-section for a semi-classical black hole at a proton–proton collider is then given by

$$\sigma^{pp}(s, x_{min}, n, M_D) = \int_0^1 2z dz \int_{\frac{(x_{min}M_D)^2}{y(z)^2 s}}^1 du \int_u^1 \frac{dv}{v}$$

$$\times F(n)\pi r_s^2(us, n, M_D) \sum_{i,j} f_i(v, Q) f_j(u/v, Q) \quad (6.5)$$

where M_D is the n-dimensional reduced Planck mass, $z = b/b_{max}$, $x_{min} = M_{BH,min}/M_D$, n is the number of extra dimensions, and $F(n)$ and $y(z)$ are the factors introduced by Eardley and Giddings (2002) and by Yoshino and Nambu Yoshino and Nambu (2003); Yoshino and Rychkov (2005) to parameterize the fact that not all of the energy of the partons is available for gravitational collapse. The $(n + 4)$-dimensional Schwarzschild radius is given by

$$r_s(us, n, M_D) = k(n)M_D^{-1}[\sqrt{us}/M_D]^{1/(1+n)} \quad (6.6)$$

where

$$k(n) = \left[2^n \sqrt{\pi}^{n-3} \frac{\Gamma((3+n)/2)}{2+n}\right]^{1/(1+n)}, \quad (6.7)$$

Furthermore, $M_{BH,min}$ is defined as the minimal value of black hole mass for which the semi-classical extrapolation can be trusted. Note that the sum is over all the partons considered in the reaction. The reduced Planck mass M_D is equal to M_* multiplied by a numerical factor which depends on the number of dimensions. The parameter M_D is bounded by phenomenological studies.

As emphasized in Chap. 5, the closed trapped surface construction is only valid in the limit where the mass of the black holes and hence the centre-of-mass energy is much larger than the scale of quantum gravity. The black hole formed in that limit is a semi-classical one. The decay of semi-classical black holes, which are thermal objects, is expected to be well described by Hawking radiation. They are thus expected to decay to many particles (see Chap. 3 for a detailed discussion). It is, however, now well understood Anchordoqui et al. (2004); Meade and Randall (2008); Calmet et al. (2008) that it is very unlikely that the LHC will see semi-classical black holes as the centre of mass energy is not high enough to produce them. This is because only a fraction of the energy of the partons is available for black hole formation Yoshino and Nambu (2003); Yoshino and Rychkov (2005) and the parton distribution functions fall off very fast as energy increases. The ratio of the lightest semi-classical black hole mass to the Planck mass is estimated to be about 5 in ADD, while it could easily be 20 for RS Meade and Randall (2008).

Besides semi-classical black holes, one expects that there are the Planckian quantum black holes introduced in Chap. 5. The energy at the LHC might be high enough to produce these holes if the Planck scale is low enough Calmet et al. (2008). From that perspective the LHC is setting the tightest limits to that on the energy scale at which quantum gravitational effects become important. These Planckian quantum black holes are defined as the quantum analogues of ordinary black holes as their mass and Schwarzschild radius approach the quantum gravity scale. They are non-thermal objects: their space-time is not semi-classical and they do not necessarily have a well-defined temperature. In many respects they are perhaps more analogous to strongly coupled resonances or bound states than to large black holes. Planckian quantum black holes presumably decay only to a few particles, each with Compton wavelength of order the size of the Planckian quantum black hole (QBH). It seems unlikely that they would decay to a much larger number of longer wavelength modes.

The cross-section for the production of Planckian quantum black holes is usually extrapolated from the semi-classical one. They are not expected to have high angular momentum. Indeed, the incoming partons are effectively objects which are extended in space-time, so their typical size is fixed by M_P^{-1}, i.e. the size of the fluctuations of space-time due to quantum gravity, which is also the interaction range of the semi-classical formation process in the limit of a Planckian quantum black hole. Thus, the impact parameter and hence the angular momentum of the QBH are expected to be small — at impact parameter M_D^{-1} the classical angular momentum would be order one at most. A classical black hole of this size with large angular momentum would have to spin at faster than the speed of light. Thus, the spin down process (see Chap. 3) before the final explosion discussed in the context of semi-classical black holes does not take place here. Quantum black holes could decay immediately to a small number of final states or they could form remnants (stable objects). Furthermore, unlike their semi-classical counterparts, Planckian quantum black holes could have quantized masses Calmet et al. (2012).

If quantum black holes decay, symmetries are useful to predict their decomposition modes. One needs to decide which symmetries are preserved by quantum gravity. Typically one assumes that gauge symmetries are preserved, while Lorentz invariance and global symmetries might be broken by quantum gravitational effects. Thus, one expects that quantum black holes can be classified according to representations of $SU(3)_c$ and that they will carry an electric charge. The formation of a Planckian quantum black hole QBH_c^q (i.e. in the c-representation of $SU(3)_c$ and with charge q) in the collision of two partons p_i, p_j has been considered in Calmet et al. (2008). A proton consists of the following partons: quarks, anti-quarks and gluons. Quarks are in the $\mathbf{3}$-representation of $SU(3)_c$, while anti-quarks and gluons are in the $\mathbf{\bar{3}}$- and $\mathbf{8}$-representations respectively. Since $SU(3)_c$ is preserved, black holes in the following representations can be produced in the collision of two protons: $\mathbf{3} \times \mathbf{\bar{3}} = \mathbf{8} + \mathbf{1}, \mathbf{3} \times \mathbf{3} = \mathbf{6} + \mathbf{\bar{3}}, \mathbf{3} \times \mathbf{8} = \mathbf{3} + \mathbf{\bar{6}} + \mathbf{15}$ and $\mathbf{8} \times \mathbf{8} = \mathbf{1}_S + \mathbf{8}_S + \mathbf{8}_A + \mathbf{10} + \mathbf{\overline{10}}_A + \mathbf{27}_S$. Thus most of the time the black holes which are created carry a $SU(3)_c$ charge and come in different representations of $SU(3)_c$. This classification allows one to make predictions for the decomposition of quantum black holes.

Several event generators have been especially designed to study the production of black holes at colliders Dai et al. (2008); Cavaglia et al. (2007); Dimopoulos and Landsberg (2001); Harris et al. (2003); Gingrich (2010). From the non-observation of microscopic black holes at the Large Hadron Collider, limits on the Planck scale in the few TeV regions have been derived. This is a very rapidly evolving field and the interested reader should read the latest ATLAS and CMS papers on the topic for up to date limits.

6.3 Microscopic Black Holes in Cosmic Ray Experiments

Microscopic black holes could also be produced in the scattering of highly energetic cosmic rays with nuclei in the Earth's atmosphere or crust. Cosmic rays with energies of 2×10^{11} GeV have been observed. Although the flux of the most energetic rays is small, new energetic particles can be produced when they collide with nuclei in the Earth's atmosphere. The centre-of-mass energy available to produce these new objects can reach some 600 TeV. In other words, one can hope to probe new physics beyond the standard model up to energies much higher than those generated at colliders.

However, the problem is to identify the new particles with detectors on Earth. In the case of microscopic black holes, two ideas have been proposed. One relies on the existence of very energetic Earth-skimming neutrinos. Within the standard model of particle physics, one expects that neutrinos can be converted to charged leptons when hitting a nucleus within the Earth's crust. If the collision happens relatively close to the surface, the charged lepton can escape the Earth and be detected. If there were a new strong interaction for neutrinos at a few TeV such as the formation of a small black hole, one would expect to see fewer of these Earth-skimming neutrino events Feng and Shapere (2002); Feng et al. (2002); Calmet and Feliciangeli (2008). In the case of Planckian quantum black holes, it has been shown Calmet et al. (2012) that the standard model background is much smaller than the expected signal. Also studying how two showers generated by the particles emitted by the quantum black hole overlap, partly resulting in oval shaped imprints, could be used to search for quantum black hole creation in the upper atmosphere.

The composition of cosmic rays is not well understood. In particular, we do not know with certainty at this point of what the most energetic components are made. They could consist of heavy nuclei and protons, as it is currently thought, but also of neutrinos. The cross-section for the production of a small black hole in a nuclei-nuclei collision has been given above in Eq. (6.5). In the case of neutrino-nuclei collisions one has:

$$\sigma(E_\nu, x_{min}, M_R) = \int_0^1 2z dz \int_{\frac{(x_{min} M_R)^2}{y(z)^2 s_{max}}}^1 dx\, F(n) \pi r_s^2(\sqrt{\hat{s}}, M_R) \sum_i f_i(x, Q)$$

(6.8)

where $M_R = M_P/\sqrt{8\pi}$ is the reduced Planck mass, $x_{min} = M_{BH}^{min}/M_R$ is the ratio of the minimal black hole mass which can be created to the reduced Planck mass, $F(n)$ is the Eardley-Giddings correction which describes the fact that not all of the energy of the partons is available for black hole formation, $y(z)$ is the inelasticity function calculated in Yoshino and Nambu (2003); Yoshino and Rychkov (2005), $\hat{s} = 2x m_N E_\nu$ where m_N is the nuclei mass, E_ν is the neutrino energy and $f_i(x, Q)$ are the parton distribution functions. The black holes produced in the reaction $\nu N \rightarrow BH$ can be charged under U(1), SU(3) but they could in principle also be neutral under these two gauge symmetries.

The number of black holes expected to be observable by a cosmic-ray experiment is given by

$$N(M_R) = N_A T \int dE_\nu \int_0^1 2z\, dz \int_{\frac{(x_{min} M_R)^2}{y(z)^2 s_{max}}}^1 dx \frac{d\Phi}{dE_\nu} A(y E_\nu) F(4) \pi r_s^2(\sqrt{\hat{s}}, M_R)$$

$$\times \sum_i f_i(x, Q)$$

(6.9)

where $\frac{d\Phi}{dE_\nu}$ is the flux of cosmic neutrinos, N_A is the Avogadro number, T is the running time of the experiment and $A(y E_\nu)$ is the acceptance of the experiment under consideration. Typically, the bounds obtained from current cosmic ray experiments are of the same order of magnitude as those obtained from collider experiments despite a higher centre-of-mass energy. There reasons are large uncertainties in the composition of the most energetic cosmic ray and of their fluxes.

6.4 Conclusions

In this chapter we discussed models with low scale quantum gravity. Such models suppose that either we live in more than four dimensions or that there is a large hidden sector of particles that only interact gravitationally with the standard model. We then discussed how colliders and cosmic ray experiments can hunt for small black holes. Remarkably the Large Hadron Collider at CERN is setting some of the strongest limits on the Planck scale.

References

Anchordoqui, L.A., Feng, J.L., Goldberg, H., Shapere, A.D.: Phys. Rev. D **65**, 124027 (2002)

Anchordoqui, L.A., Feng, J.L., Goldberg, H., Shapere, A.D.: Phys. Rev. D **68**, 104025 (2003)

Anchordoqui, L.A., Feng, J.L., Goldberg, H., Shapere, A.D.: Phys. Lett. B **594**, 363–367 (2004)

Antoniadis, I., Arkani-Hamed, N., Dimopoulos, S., Dvali, G.R.: Phys. Lett. B **436**, 257–263 (1998)

Arkani-Hamed, N., Dimopoulos, S., Dvali, G.R.: Phys. Lett. B **429**, 263–272 (1998)

Banks, T., Fischler, W.: hep-th/9906038 (1999)

Beringer J. et al., [Particle Data Group Collaboration], Phys. Rev. D **86**, 010001 (2012), see also their website: http://pdg.lbl.gov/ for up-to-date information

Calmet, X., Fragkakis, D., Gausmann, N.: Non-thermal small black holes. In Bauer, A.J., Eiffel, D.G. (eds.) Black Holes: Evolution, Theory and Thermodynamics, pp. 165–170. Nova Science Publishers, New York, 2012

Calmet, X., Gong, W., Hsu, S.D.H.: Phys. Lett. B **668**, 20–23 (2008)

Calmet, X., Feliciangeli, M.: Phys. Rev. D **78**, 067702 (2008)

Calmet, X., Hsu, S.D.H., Reeb, D.: Phys. Rev. D **77**(125015), 1 (2008)

Calmet, X., Caramete, L.I., Micu, O.: JHEP **1211**, 104 (2012)

Cavaglia, M., Godang, R., Cremaldi, L., Summers, D.: Comput. Phys. Commun. **177**, 506–517 (2007)

Dai, D.C., Starkman, G., Stojkovic, D., Issever, C., Rizvi, E., Tseng, J.: Phys. Rev. D **77**, 076007 (2008)

Dimopoulos, S., Landsberg, G.: Proceeding International Workshop on Future of Particle Physics (Snowmass) (Preprint SNOWMASS-2001-P321)

Dimopoulos, S., Landsberg, G.L.: Phys. Rev. Lett. **87**, 161602 (2001)

Eardley, D.M., Giddings, S.B.: Phys. Rev. D **66**, 044011 (2002)

Feng, J.L., Shapere, A.D.: Phys. Rev. Lett. **88**, 021303 (2002)

Feng, J.L., Fisher, P., Wilczek, F., Yu, T.M.: Phys. Rev. Lett. **88**, 161102 (2002)

Giddings, S.B., Thomas, S.D.: Phys. Rev. D **65**(5), 056010 (2002)

Gingrich, D.M.: Comput. Phys. Commun. **181**, 1917–1924 (2010)

Gogberashvili, M.: Int. J. Mod. Phys. D **11**, 1635–1638 (2002)

Harris, C.M., Richardson, P., Webber, B.R.: JHEP **0308**, 033 (2003)

Huber, S.J.: Nucl. Phys. B **666**, 269–288 (2003)

Kabat, D.N.: Nucl. Phys. B **453**, 281–302 (1995)

Larsen, F., Wilczek, F.: Nucl. Phys. B **458**, 249–266 (1996)

Meade, P., Randall, L.: JHEP **0805**, 003 (2008)

Randall, L., Sundrum, R.: Phys. Rev. Lett. **83**, 3370–3373 (1999)

Vassilevich, D.V.: Phys. Rev. D **52**, 999–1010 (1995)

Yoshino, H., Rychkov, V. S.: Phys. Rev. D 71, 104028 (2005) [Erratum-ibid. D 77, 089905 (2008)]

Yoshino, H., Nambu, Y.: Phys. Rev. D **67**, 024009 (2003)

Chapter 7
Conclusions

Although black holes provide a unique probe of gravity in its most extreme regime, they are not just a laboratory for exploring classical general relativity. Their quantum effects are also important and in this book we have presented a brief state-of-the-art overview of our current understanding of these effects. In the absence of a final theory of quantum gravity, we have seen that understanding the quantum properties of black holes provides a glimpse of the elusive laws of fundamental physics underpinning the Universe.

Our story has involved a combination of mathematical and physical discourse and touched upon both theoretical and observational developments. We began with a presentation of the well-known properties of classical black holes in general relativity, stressing the links between black hole mechanics and thermodynamics. We focussed mainly on the Schwarzschild and Kerr solutions but also described their higher-dimensional counterparts.

We then encountered the simplest and best understood type of black hole quantum effect: particle creation. This depends on the use of quantum field theory in the curved space-time background of the black hole solution, the prediction of Hawking radiation being the most important result of this approach. For later applications, we described the emission in some detail (including grey-body factors) and we also touched briefly on the information loss paradox, though without discussing its possible resolutions.

The only astrophysical realization of these effects involves primordial black holes, so this was our next topic. Quantum effects play a double role in this context since both the formation and evaporation of such black holes are a consequence of quantum fluctuations. Depending on their mass, which determines their lifetime, primordial black holes could have a huge variety of cosmological effects and this makes them a unique probe of the early universe. Even if they never formed, the upper limit on the fraction of the universe going into them as a function of mass provides important constraints on models such as inflation. More positively, black holes evaporating at the present epoch might explain certain anomalies associated with cosmic rays and dark matter.

X. Calmet et al., *Quantum Black Holes*, SpringerBriefs in Physics,
DOI: 10.1007/978-3-642-38939-9_7, © The Author(s) 2014

We then returned to more formal considerations and discussed how black holes can form in the collision of two particles. The Eardley and Giddings construction elegantly demonstrates that even for non-zero impact parameter, classical black holes do form in the ultra-energetic collision of particles. This construction can be extended into the semi-classical regime using a path integral formulation.

Finally, moving into more speculative domains, we considered the formation and decay of Planckian quantum black holes. Since quantum gravitational effects should be important here, these are in a sense the most quantum black holes of all. We described how modifications of general relativity could lead to quantum gravitational effects at a few TeV, leading to the production of microscopic black holes by the LHC or cosmic rays. At the time of writing, there is still no sign of physics beyond the standard model of particle physics. However, the LHC is setting the tightest limits to date on the Planck scale. These are evolving constantly, so we refer the reader to the literature for the most up-to-date developments.

Black holes have been intensively researched for over 50 years and there have been many astounding discoveries, both theoretical and observational. Nevertheless, there is much about black hole physics which is still poorly understood. In this brief book, we have tried to present a flavour of some current research on the quantum properties of black holes. We hope that we have conveyed a sense of where the field currently stands, and what the outstanding open problems are, although there is much that we have had to leave out due to the restrictions of space. With the LHC running at higher energies, more detailed cosmological data and new theoretical developments, we anticipate exciting years ahead for black hole physics.

April 2013

Index

X. Calmet et al., *Quantum Black Holes*, SpringerBriefs in Physics
DOI: 10.1007/978-3-642-38939-9, © The Author(s) 2014